BEHAVIOLOGY REVOLUTION IN PSYCHOLOGY

Why we do what we do?

WHAT HAPPENS IN OUR MINDS WHEN WE INITIATE MOTIONS?

This book is a modified title of Behaviology

By: Scientist and Researcher: Eddie Rafii

ISBN: 9798714114151
ISBN-10: 1979653178
Library of Congress Control Number: 20209206

PRESENTATION OF THE BOOK

WE DO NOT MOVE UNLESS WE ARE BOTHERED, every single human motion is caused by a bothersome. We remain motionless if nothing bothers us.

What would be the first
motion of your body if
nothing bothered you?

WHAT ARE MORALITIES?

Conscience

You are the guardian of an orphan, and you will get a bad feeling if you spend his or her money on yourself. Does morality stop you from stealing the money, or does the fear of future suffering prevent you from the dishonest action?

Punctuality

I suffer when I am late for my appointments. Does morality or the fear of suffering force me to be punctual?

ARE THE MORALITIES HUMAN VALUES?
OR
ARE MORALITIES NOTHING BUT FEARS OF SUFFERING?

Our actions and reactions are based on our balance

We perform motions only when we lose our balance, and we remain motionless as long as we don't get imbalanced

· ·

Ultimate comfort inactivates us.

Problems, troubles, and discomforts activate us.

We remain motionless if nothing bothers us.

Are the following attributes bad or good?
- ➤ Greed
- ➤ Supremacy
- ➤ Aggression

How would our lives be today if we had removed the above traits from Steve Jobs and Bill Gates' characters when they were twenty years old?

It is so obvious that we would not have such great improvements in computer and communication technologies if we had removed these attributes from Steve Jobs and Bill Gates' characters.

Talents are grown by curiosity and poverty.

Technologies are grown by greed, supremacy, and aggression.

If you ever tried to feed pigeons from your hand, you have seen that they come around you, but are afraid to get too close to you. Here, there is a *want* to eat and a *fear* of being caught by you. The sum of the outcome of want and fear determines pigeon motions.

Want to eat and *fear of* being caught

A hungry person who needs food goes to a market but doesn't have enough money to pay for the food. Here, there is a *want* to eat and a *fear* of being caught for stealing food. The person makes motion according to the sum of the outcomes of his want and fear, he won't steal if his fear is greater than his want, but he will steal if his want is greater than the fear.

The mechanism and functionality of humans' and animals' motions are identical.

Evaluating characteristics as just positive or negative traits is not a proper way to determine one's qualification. For example, we judge people as dishonest or honest. A dishonest person cannot be trusted at all (zero trust) and an honest person can be trusted one hundred percent. Dividing people into only two groups creates an incorrect evaluation, however. the degree of trust could vary from 0 to 100, but each person has a certain degree of trust. There is no tool available to measure the degree of a characteristic to determine the property of a person, but we can evaluate roughly the degree of each character in a person from 0 to 100.

There is nothing beyond our bodies such as psyche

Emotions and feelings are just chemical processes in the brain

Mental disorder is an improper term and must be replaced with chemical disorder

■■■

THINK NEUTRALLY

Some of our personal beliefs that have no logical or scientific basis prevent us from seeing reality. We cannot think neutrally when our minds are filled with untrue beliefs. Most of our beliefs—especially those that were placed in our minds when we were children—have no logical or scientific basis. We are often unaware of how they interfere with our ability to face new ideologies that differ from our beliefs.

This book is a guide to helping you see reality based on neutral thinking. It describes the step-by-step method of simple logic to provide you with the tools to dig through life's aspects and see the deep secrets of human nature. In doing so, you will be enlightened by a new vision by the time you finish reading the book. This new vision will make it possible to analyze your unanswered questions and to find answers.

One of the most difficult challenges in life is to erase all beliefs, which involves clearing and preparing our minds for accurate information.

While reading this book, you will develop a new way of thinking. However, if you find yourself having difficulty accepting the book's points, it may be that untruthful beliefs have been rooted so strongly in your mind that it will become the ultimate challenge to replace them. When beliefs are deeply embedded in the mind, you may discover that it is practically impossible to think purely logically. After all, altering our original beliefs to some other form is no simple task. Therefore, this book will help particularly those who initially have an open mind and want to seek life's truth.

PRAISE THE BOOK BY;

A. Farnoody, PhD, psychology Jondishapoor University.

INCREDIBLE DISCOVERIES OF HUMAN NATURE

There are many books about inner daily life, but there are not many that combine real originality with intellectual integrity. But it is exactly this combination that Eddie Rafii has been able to produce. The author has given us a careful study that will be valued for years to come. Though the present book demonstrates indebtedness to the classics, it is not a book about them; it represents, instead, genuinely original work.

What strikes me at once is the comprehensiveness of the current undertaking. Many contemporary books deal with particular aspects of inner life, but this one is different in that it deals with an astonishing variety of important topics; much of its freshness of treatment arises from a lifetime's thinking, observing, and experimenting. The author has undertaken to examine a wide spectrum of experiences, from simplicity to multiplicity, from joy to suffering, and vice versa.

Since the finished product is the outcome of wide

reading and careful thinking, it is not the sort of book that can be dashed off quickly. The literary style of the author is clear and simple, which is especially valuable since the subject is not simple. In a nutshell, each chapter is designed to tell the truth.

The principles the author uses—rooted in ancient wisdom—are made astonishingly contemporary. The author understands very well that emphasis upon simplicity may itself become a snare. But a genuine cultural revolution would ensue if considerable numbers were aware of their daily behavioral activities.

The greatest problems of our time are not technological, for these, we handle fairly well. They are not even political or economic because the difficulties in these areas, glaring as they may be, are largely derivative. The greatest problem is unhappiness; unless we can make some progress in this realm, we may not even survive—this is how advanced cultures have declined in the past. For this reason, I can become a truly mature being by cultivating a life in the manner that Eddie Rafii describes.

FOREWORD

This book is the result of long-term observations and experiments, a manual for achieving a more serene life in the midst of the constant stress of daily life. The book draws on common sense, everyday examples to help ordinary people improve their lives by embracing problems, rather than fleeing from them. By following the step-by-step procedure, readers will develop greater patience, problem-solving skills, and general happiness.

Currently, so-called happiness studies are in vogue—psychologists and sociologists are publishing research about how citizens of developed countries often are unhappy in the midst of relative affluence. This book offers nonacademic readers practical ways to increase contentment while also developing their thinking skills by respectfully challenging their belief systems. This book provides the critical tools to dig through readers' everyday experiences in order to understand human nature, which in turn sheds light on philosophical problems such as free will. As a result, readers become more psychologically balanced so that they are more patient and less likely to be disturbed by others' anger. In addition, readers learn how to enjoy life by appreciating what they already have, rather than suffering from society's expectations of what they should acquire.

Jeff Karon

I would make an analogy between a car with a broken water pump and a short-tempered person. If you have a used car and you drive uphill, you expect the radiator to boil because of the broken water pump; thus, you don't get mad at the car when it is boiling. If your husband is short-tempered, and you have the knowledge that the related gene makes him boil easily without his control, then any time he boils, you handle him as if he had no control over his reaction.

About Eddie Rafii

Eddie Rafii is one of the world's rare persons; he has never read a psychology book but his discoveries have revolutionized psychology. Eddie is a researcher and scientist, a deep observer and thinker, a genius from Persia. He has spent most of his life finding the cause of human behaviors. He has discovered seven facts in human nature that are the basis of our behaviors. **"WE DO NOT MOVE UNLESS WE ARE BOTHERED,"** one of his remarkable discoveries, opened a new window into the world of psychology.

Eddie says, "I'm an ordinary man, but I was able to alter my train of thought and beliefs to see life's realities." This was no easy task; it took him seven years to eliminate all cultural, religious, and other beliefs that were strongly instilled and embedded in his mind.

At age nineteen, he was consumed with a thought-provoking question: What is the cause of an action? What happens in our minds and bodies that causes us to initiate motions and subsequent movements as we sit doing nothing (remaining immobile)? To answer this question, Eddie conducted various experiments and observations on himself and others. After extensive research, he discovered a clue that logically answered his initial question as well as several related questions. As a result, all his confusion and conflict were resolved.

For seven years, he fought the battle between his beliefs of origin and his sense of logic until he successfully overturned the untrue beliefs and cleared his mind to complete his thoughts. Finally, over the last several years, Eddie has been motivated to organize his thoughts into this book and share them with others. This book is the result of a lifetime of thinking, observing, and experimenting.

Eddie is multitalented, he is not only a researcher but he is also an inventor. He never studied mechanical engineering but he has invented the Rockless Table.

At an early age, Eddie graduated from the University of Tabriz with a master's degree in agricultural engineering and then earned a diploma from the Industrial Management Institute, Course of Studies of Societe ACTIM in France. He also is a certified graduate of Centro Studi Agricoli Shell in Italy and has conducted experimental research in vineyards and wineries in Romania. He immigrated to the United States in 1978. His real name is Yadollah Rafii Neishaboori, but he simplified it to Eddie Rafii in the American tradition when he became a U.S. citizen.

DO NOT READ RANDOMLY

It's highly recommended that you read this book from the beginning because it employs a chain of logic, and each subject builds on the conclusions of previous subjects. You will get the wrong impression if you choose to read sections randomly. As an example, you might think that it doesn't make sense to say, "A sheet of paper will reach to the moon if you fold it fifty times," but it would be very easy to understand when you study the basic mathematics involved.

TABLE OF CONTENTS

TABLE OF CONTENTS

Eddie Rafii discovered seven major facts in human nature underlying every single human motion:

Fact 1- We always want to be in balance.

Fact 2- We make motions only when we lose our balance.

Fact 3- Loss of balance always generates want or fear in us.

Fact 4- Suffering and joy always accompany every motion.

Fact 5- More suffering leads to more enjoyment.

Fact 6-The sum of the outcome of our want and fear determines our motion.

Fact 7-The degrees of our personality and physical characteristics, and mental and environmental conditions, determine the levels of our enjoyments and sufferings, and the sum of the outcomes of those enjoyments and sufferings determine our motions.
So, what does that say about human nature?

Chapter 1

CAUSE OF MOTIONS

What would be your first motion if nothing bothered you?
We do not move unless we are bothered. Every single human motion is caused by a bothersome. We remain motionless if nothing bothers us.

Imagine sitting on the sofa in your living room, completely comfortable, with nothing at all to bother you. You are not hungry, thirsty, or tired. You are not cold, hot, or sick. You are not worried about bills coming due or your job or business. You are not bored or sleepy. The telephone does not ring, and absolutely nothing makes you uncomfortable, worried, curious, excited, or upset. **Would you make any movement if nothing bothered you?**

Of course, such moments are rare in our lives, and if they occur, they don't last very long. For example, when you wake up in the morning on your day off, and you have nothing to do at that time, continue to stay in bed for a while with no movement.

If you continue the above experiment, you will observe that sooner or later, your body generates some motion, such as:

(1) You change your sitting position.

(2) You scratch your forehead.

(3) You blink.

(4) You go to the refrigerator to get a drink.

(5) You turn on the heater.

Why do these motions happen? Let's look more deeply at these situations to find the cause of these motions:

(1) When sitting on your sofa for a while, your body becomes tired; consequently, you begin to feel uncomfortable. Your body generates motion to change its sitting position to regain your comfort. Therefore, tiredness bothered you, leading your body to generate motion.

(2) An itchy forehead bothered you. You scratched it to regain your comfort.

(3) You blinked because your eyes were dry or irritated, which made you feel uncomfortable. Your body carried out a motion (eyes blinking) to regain your comfort.

(4) You went to the refrigerator to get a drink. You were affected by an internal factor—thirst. You lost your comfort (balance), so you moved to

get a drink to regain your balance.

(5) You turned on the heater because the room was getting cold. The cold, an external factor, affected you, causing an imbalance on you; thus, you turned on the heater to regain your balance.

There is always some factor that affects you, causing an imbalance in you, and you make motions to regain the balance.

If such factors (tiredness, itching, dry eyes, thirst, and cold) did not exist or did not bother you, you would never move. In other words, **WE DO NOT MOVE UNLESS WE ARE BOTHERED**. [F2]

If we are sitting completely unbothered and therefore not moving at all, are we alive? To many, it may look like we are dead, but clearly, we are not. Internally, our bodies are moving; for example, our blood and heart are moving. Is the movement essential to life? When we see a four-year-old boy running, playing, and never sitting down, we say he is lively, thus making an unwitting comparison to a person who demonstrates less activity. Hence, the more activity we have, the more alive we are.

Our actions and reactions are caused by bothersome factors. Troubles and problems generate discomfort (suffering) in us, we perform motions to regain our comfort (enjoyment).

Suffering and joy always accompany every motion. If we want more enjoyment, we must seek factors that bother us. We need more trouble to be more alive. Cold, heat, hunger, thirst, jealousy, tiredness, sickness, debt, poverty and so on makes us to lose our comfort (imbalance), we move to regain our comfort (balance). These movements and activities make us alive. Are the cold, heat, hunger, thirst, jealousy, tiredness, sickness, debt, and poverty bad or there are necessary for life? To find the answer, let's go back to the previous claim that *we do not move unless we are bothered* and try to find an example to prove otherwise.

The next time you go to a zoo, take a moment to examine old lions. Older lions tend to sit somewhere in their cages or under a tree for a long time, with little or no motion. Their movements are limited to blinking or slight movements of their heads. They only move if they are hungry or when they get tired of sitting. Why don't they have any other motions? They are simply not bothered by any other factor.

You may have observed the same in old cats. Younger cats are more active because certain factors make them lose balance; however, they do not affect older cats. For instance, if you move a toy in front of a kitten, it becomes excited (loses balance) and makes a motion by catching the toy, but if you move the toy in front of an old cat, you rarely see any reaction.

The same phenomenon occurs in humans. The factors that make the young lose their balance generate no excitement in older people. When you see older people sitting without movement on a couch for a long time, it is because nothing bothers them. Therefore, *we remain motionless if nothing bothers us.* Further thought experiments lead us to seven basic facts:

Fact one: We always want to be comfortable (what we will call *balance*). [F1]

Fact two: We make motions only when we lose our balance. [F2]

Fact three: A loss of balance always generates a *want* or *fear* in us. [F3] For instance, when we get thirsty, we lose our comfort and balance, so we *want* to drink. We lose our balance when it gets cold, which creates *fear* of catching a cold.

Loss of balance by seeing a delicious cake generates a *want* to eat

Fact four: Suffering and joy always accompany every motion. [F4] For example, when our forehead itches, we feel uncomfortable, so we suffer a little and then enjoy it when we scratch it.

Fact five: More suffering leads to more enjoyment. [F5] For instance, if we drink as soon as we get thirsty, we suffer and enjoy a little; but, if we have no water for hours, we suffer more and enjoy it more once we do drink water.

Before we review facts six and seven, take a moment and see if you can find examples in your daily motions that contradict these facts. Doing so is very important because these facts are the basis for additional topics I discuss in this book. New conclusions will change our views completely.

I wish I could discuss these subjects face-to-face with every single reader of this book so that I could listen to and analyze each person's objections. Since face-to-face discussion is impossible, I analyzed several common examples in chapter three to address common objections that you may have in your mind. A careful study of chapter three will clear possible contradictions in your mind.

The experiments and examples mentioned in this chapter show us that every single human motion is caused by a bothersome, and we remain motionless if nothing bothers us.

Conclusion:
We do not move unless we are bothered. [F2]

Note:
You will read 28 facts in this book that are marked like [F1] to [F28]. These facts are the new discoveries in human nature.

Chapter 2

ADVANCED VISION

BASIS OF HUMAN BEHAVIOR

Every person does hundreds of motions every day; a single individual does millions of actions throughout his or her life. Let's take a look at a day in the life of an individual. You may look at your own life if it is easier.

You wake up in the morning. Before you think of anything or any factor affecting you, you are in balance (comfortable), but this lasts for a short time.

You get up as soon as you remember that you either have an appointment or need to go to work or school. What happened?

You lost your balance, so you start moving to regain it, but you still haven't regained it completely. You need to do other movements to get to work or school. Once you get to your destination, you will regain your balance. Let's see what your next motions will be after you get up.

You have to urinate, so you lose your balance. You use the restroom to regain your balance. You lose your balance because you feel dirty or not fresh, so you take a shower to regain your balance. You see one of your family members and say good morning. What would happen if you didn't say good morning? You'd feel uncomfortable, so you say good morning to regain your balance. Of course, some people don't say good morning and don't feel imbalanced, because such forms of communication partially depend on society and our upbringing. A person who sees a family member and does not feel uncomfortable for not saying good morning doesn't lose his or her balance and consequently makes no motions.

Let's go back to the motions that you do on a normal day. After you've said good morning, you see a used napkin on the floor. You lose your balance, so you pick it up and put it in the trash to regain your balance. What would happen if you didn't pick the napkin up? You would feel bad every time you saw the napkin, and it would bother you until you picked it up and put it in the trash can.

You put on your clothes to go to work. What would happen if you went to work in your pajamas? Of course, you would feel uncomfortable. When you are at home, and there is no one to see you in your pajamas. What factor makes you lose your balance, pushing you to put on your clothes? In this case, you lose your balance because of an incident that will occur in the future. You think that if you were to go to work in your pajamas, you would lose your balance. We do many motions daily to prevent our discomfort and imbalance in the future, like when we go to school to live a better life in the future, when we go on a diet to avoid gaining weight or being unhealthy, or when we save money to prevent ourselves from having a hard time in the future. Such factors make us do things because of a fear of losing our balance and suffering in the future. When we go to school, we fear leading a difficult life in the future. When we go on a diet, we fear losing our balance in the future due to weight gain or bad health. When we save money, we fear having a hard time in the future. The chapter on balance further expands on this topic.

Let us consider our daily motions again. You arrive at work, and your supervisor looks at his watch, indicating that you are late. You lose your balance and apologize to him to regain your balance. If you feel that apologizing is not enough to regain your balance, you may explain why you were late, perhaps even lying about the terrible traffic. We sometimes lie to regain our balance. Although you make excuses to regain your balance, you may not completely regain it, and your body may remain a little underbalanced. In this scenario, you may feel that your supervisor didn't believe your excuses. You go to your desk and see a pile of work, so your balance falls to a lower level.

The air conditioning system in the office is broken today, so your balance falls even lower. Then on the corner of the desk, you see an envelope with your name on it. You lose your balance and feel that you must open it. If you don't open it, you'll remain imbalanced until you open it. Once you open it, you see that it's the promotion you've always wanted, and suddenly your balance jumps from a low level to a very high level.

I wish I could make a gauge to measure the level of balance, like a thermometer that determines our body's temperature. I also wish that I could make another tool to measure how much we suffer and enjoy. With such tools, it would be very easy to analyze what happens to us regarding each motion.

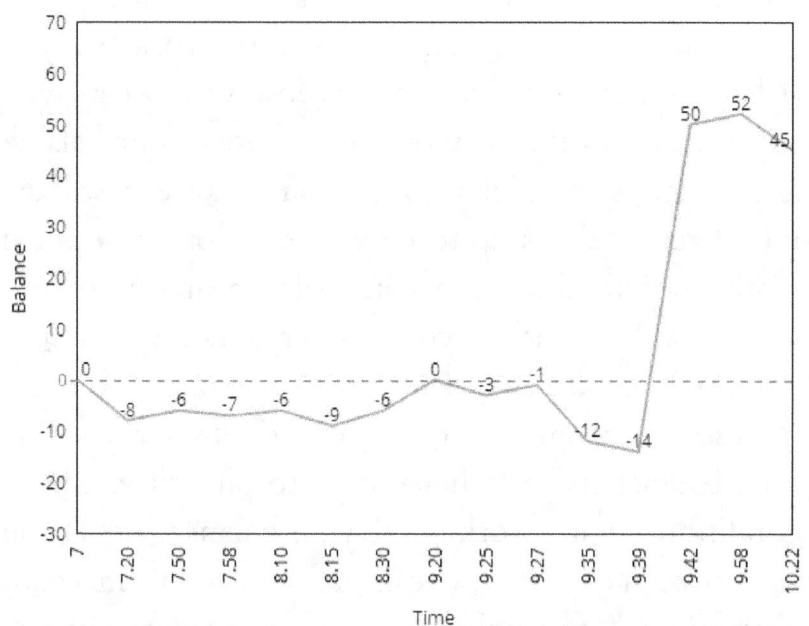

Imagine that we have a gauge called the Balance Gauge. In the morning, we attach it to you and observe what will happen. When you wake up, before any thoughts enter your mind, you are relaxed and comfortable, so the gauge shows zero. Then you remember that you have to go to work. The gauge shifts to minus eight. You get up to get ready, and the gauge changes to minus six. You feel that you have to use the restroom and lose your balance a little more, so the gauge shifts back to minus seven. After using the restroom, the gauge goes back to minus six. You look in the mirror, and you hardly recognize your face. You lose your balance, and the gauge changes to minus nine. You put on your makeup or shave, and your balance goes back to minus six. You get to work, so your balance goes back to zero. However, when your supervisor sees that you're late, you lose your balance, and it drops to minus three. You make excuses, and your balance shifts up to minus one. You see a pile of work on your desk, and your balance drops down to minus twelve. The air conditioner is not working, so your balance drops even lower—to minus fourteen. You see the promotion on the corner of your desk, and your balance makes a huge jump to plus fifty. The air conditioner starts working, so your balance goes to plus fifty-two. Your spouse calls and says the mechanic charged $300 to fix the brakes on the car; your balance shifts down to plus forty-five.

Now imagine that on the same day, we attached a second tool to you to measure joy and suffering. Let's call this the Pleasure Meter.

We have set the meter to zero and attached it to you before you wake up. You wake up, and as soon as you remember that you have to go to work, you lose your balance and begin to suffer; the pleasure meter counts down and stops at minus five. You suffered five units. When you start getting up, the pleasure meter starts going up and stops when you get up. For instance, it stops at minus three, indicating that your enjoyment improved by two units. Then, if you feel you have to use the restroom, the meter starts counting down until you get to the bathroom. The meter shows minus five just before you pee. As soon as you start peeing, you feel enjoyment. The pleasure meter shows minus four and, by the time you finish, it reaches minus three. Here, you suffered two units from losing your balance and enjoyed two units when you regained your balance. You see yourself in the mirror, and you lose your balance. The meter drops to minus four. You put on your makeup or shave, and the meter shifts up to minus three. You suffered one unit from losing your balance and enjoyed one unit after completing your morning makeup. You are driving to work, and the meter keeps working until it shows minus two and then minus one. When you arrive at work, the meter is at zero.

When you first woke up in the morning and remembered that you had to go to work, you lost your balance and suffered minus five units. Your body made motions to get your balance back. The first motion was getting out of bed, and the second series of motions involved actions necessary to get to work. The total enjoyment was plus five, which balanced the total suffering from losing your balance when you remembered that you had to go to work.

If you follow the functions of the pleasure meter, you will see that it drops when you're suffering and rises when you're experiencing joy. At the end of the day, when it's time to go to bed, the meter shows a number. This number could be negative, zero, or positive.

If the meter shows a zero, it means that your total suffering was equal to your total enjoyment on that day. If the meter shows a negative number, it means that you had a bad day overall and that your total enjoyment was less than your total suffering. If the meter shows a positive number, it means that you had a good day overall, and the total enjoyment was greater than the total suffering. The number of joy units on the scale of the Pleasure Meter indicates the level of your happiness on that day. The number of the units of suffering indicates the level of your unhappiness on that day.

The Balance Gauge will reset to zero every morning when you wake up before anything comes to your mind, but the Pleasure Meter will show the same number as when you went to bad, although it will be affected by the joy of resting through the night. When you wake up in the morning—and as soon as you remember that you got a promotion—your balance on the gauge jumps to a high number, and the Pleasure Meter goes up. The Pleasure Meter stops moving upward any time other factors (such as remembering that you have to go to work) interrupt your mind from thinking about the promotion.

You start a new day with similar motions. Here, I would like to explore the question: What is life? If someone asks you to describe life in one sentence, what would your description be?

When we analyze a single day, we see an imbalance and a rebalance or suffering and enjoyment for every motion. Thus, if we look at a day in our life, from the morning when we wake up until the night when we fall asleep, suffering and enjoyment are recurring.

Can we say that life is nothing but the recurrence of suffering and enjoyment? If this is true, then what is happiness? We will analyze this in detail in the chapter on Happiness.

Regarding the first five facts outlined in chapter one, you can try to find examples that prove these facts are not correct. Most examples you might have will probably fit in one of the objections in the next chapter, and if they do not, analyze them yourself after reading this chapter carefully. Your objections must be based on real examples, as they are easier to analyze if you take them from things that have actually happened in your life. Look at your motions yesterday or today.

Reflect on just one of those motions and ask yourself what your feeling would be if you prevent yourself from doing them. For example, if you saw a poor man begging for money, and you gave him some money, imagine what your feeling would be if you hadn't given money to the beggar.

When you saw the poor man, you lost your balance. If you walked away, you would have a bad feeling and imbalance, so you gave him money to regain your balance. Of course, some people don't feel bad when they see a poor man; obviously, they won't perform any motions if they don't lose their balance. These people may think that this man is dishonest and will use money recklessly or foolishly, or they may be insensitive (indifferent or inconsiderate rather than tough-minded). When you prevent yourself from carrying out an actual motion, this method of thinking about your feelings will allow you to analyze the situation to find the true answers.

However, if you find an example that could prove the five facts false, please send me an e-mail, and I'll analyze it with you.

Conclusion:

Our actions and reactions are based on our balance. We perform motions only when we

lose our balance and we remain motionless as long as we don't get imbalanced. [F2]

She will sit there forever with no movement
if nothing bothers her

Chapter 3

CONTRADICTORIES

Note:

Before getting into the contradictories, I would like to bring to your attention that there are sufferings and enjoyments in our daily lives that are not involved with our actions such as the suffering of the loss of our favorite team in a basketball game or suffering of seeing the dead people in an earthquake on the news, or the enjoyment of watching a comedy movie or the enjoyment of seeing beautiful flowers or suffering of seeing a dead bird when we are walking in a park or the enjoyment of seeing that your friend bought a new car. We do not take any action when this kind of joy and suffering occur, these types of sufferings and enjoyments may not be counteracted in the same day but a large number of such sufferings and enjoyments, in normal life will be counteracted in a longer period of time. We will discuss suffering and joy without our motion in detail in the chapter on happiness.

As I mentioned in chapter one, I wish I could discuss these subjects face-to-face with every single reader of this book so that I could listen to and analyze your objections. Since the face-to-face discussion is impossible, I'll analyze the following common examples that will cover the objections that you may have in your mind.

Before we talk about possible contradictory examples, I would like to bring to your attention the fact that usually, action is a result of first suffering and then enjoyment; in some cases, though, we first enjoy and then suffer. For example, when we enjoy a weekend of rest at home or with friends, we'll be lazy and unhappy on the following Monday. The same can be said for returning from a vacation—we won't feel happy for a while. When some people indulge in alcoholic beverages, they feel poorly the following day. In these and other cases, we derive immediate enjoyment from a motion and suffer in the future. For example, children often prefer playing or spending time with their friends instead of doing homework. Of course, these kids will not receive a good education and will suffer because of it by having a harder life for a while in the future.

Now let's go back to some possible objections you might have.

Objection one:
"I have sex and enjoy. Where is the suffering?"

When we just take a quick look at the process, it

seems that we just enjoy sex without suffering. But we would reach a different conclusion if we analyze the situation carefully.

The sex usually starts by viewing the opposite sex's gender; we lose our balance and we want to touch (or to be touched) and kiss. We lose our balance even more by touching and kissing. To regain the balance, we perform a chain of motions to get to orgasm.

Each motion in the whole process contains suffering and enjoyment. For instance, we see the body, we lose our balance (Suffer), we perform a motion by touching the body; as a result, we enjoy some but suffer more by placing ourselves in a higher imbalance—we must perform other motions to get to orgasm and regain our balance.

Let's take a look at a porno movie to have a better understanding. Focus on impressions and the voice of a woman before orgasm. Are the impressions the signs of enjoyment? Although the actors are faking, they are trying to show us the impressions of a real scene. The enjoyment could be found only at the end by the orgasm.

We all have had the experience of being interrupted during sex. We feel unhappy and angry. Aren't these feelings signs of suffering? So having sex is not just enjoyment—there are also some hidden sufferings. One of the major causes of losing interest in having sex with our regular sex partner is that we suffer periodically by not reaching orgasm.

When we analyze the outcome of enjoyment of sex, we will observe that this enjoyment could be divided into three portions. The first portion pays off earlier sufferings of the tensions of daily life, particularly in males. The second portion pays off all the sufferings during sex, and the third portion places our body in a higher balance (relaxed) that would be counteracted by the next tension.

Objection two:

"When I hear something sad, I cry. I suffer without any enjoyment."

When we hear something sad, we lose our balance, and we want to do something to regain our balance. Sometimes when we hear something sad, we can't make any motions. For instance, when we hear that one of our loved ones has died, we can't get his or her life back, so we cry to regain our balance. After crying, we are still sad. We cry more tomorrow and the next day, but where is the enjoyment?

When we meet a person for the first time, there is no relationship and love; but when the relationship begins, we start to build it by giving and receiving until it becomes stronger and stronger. Here is enjoyment for which we never suffer. We enjoy receiving the pleasures of love and companionship, as well as the materialistic joy from flowers, gifts, or loans. These enjoyments build until the day we have to counteract them, so when we suffer from the loss of a friend or family member, we are "paying off" the enjoyment that we had before.

Sometimes when we hear something sad, we don't cry. Yet our body still makes a motion to regain its balance; for example, when we hear that a friend is sick, we take him to the hospital. We find him a good doctor or give him a room in our house, so we can take care of him. This way we try to regain our balance. Of course, taking a person to the hospital, finding a doctor, and taking care of someone in your house is not fun and involves suffering. These sufferings are the payoff from past companionship joys.

Objection three:

"I was sitting on my sofa and nothing was bothering me. I turned on my stereo, and I enjoyed music without suffering."

Before you turned on your stereo, you were bored and consequently lost your comfort, so you suffered a little. You may have remembered your favorite song, which also made you lose your balance, so you turned on your stereo. (How would you feel if you remembered your favorite song and had prevented

25

yourself from turning on the stereo? Of course, you would suffer a little.) When you enjoy music, the enjoyment is in two portions. One portion is paying off the previous suffering while the other is to set your balance at a higher level that will be adjusted by the next suffering.

Objection four:

"When I see a movie, I enjoy without suffering."

I saw you when you left the theater after viewing *Titanic*. You were so sad. Did you suffer?

"No."

"So, you enjoyed it?"

"Yes."

"Then why did you cry?"

"Because it was sad."

"So, you suffered?"

"Yes, no, no. Yes, I suffered."

"If you suffered, why did you see the movie again?"

"I don't know."

Let's analyze what happens to us when we see a movie. First of all, before we watch a movie, some factor affects us, and we lose our balance, which generates a want in us to see the movie. Sometimes this occurs from hearing others praise a new movie currently playing; we lose our balance, and we *want* to see the movie. Often the factor is our boredom, which causes us to lose our balance; we go to the movies in order to regain our balance. In either case, we first suffer and then enjoy. Most of the time, by viewing a movie, we suffer a little and enjoy more by the end of a movie (when a movie has a happy ending); however, sometimes the suffering is more than the enjoyment (when a movie has a sad ending). In this case, our body remains unbalanced for a while, which we will discuss later.

Before we analyze what happens when we see a movie, let's look at the role of the writers and directors. At the beginning of most movies, the director introduces an actor through specific scenes meant to create a desire in the audience to like the actor. When we start liking the actor, we empathize with him or her even if he or she is a bad guy or thief—we still want to see him or her succeed. An example is Robert De Niro playing the role of a professional thief in the movie *The Score.*

At the beginning of the movie *Titanic*, we see a woman, her fiancé, and another young man boarding the ship. At this time, we take no sides. As the movie continues, we see that the fiancé' is a bad guy, and the woman is innocent. We begin to like her and take her side. When the other young man comes to save her life, we like him and take his side as well. From this point forward, we are the fiancé's enemy, and we want him to lose. The first scene that affects us is when the woman wants to attempt suicide. At that moment, we lose our balance; we become worried and start suffering. Then the young man comes and saves her, which restores our balance and enjoyment. Ultimately, the movie provides a series of cycles of suffering and enjoyment by creating danger for the main characters through the devices of the fiancé, his bodyguard, or the sinking ship. When we see the main characters in danger, we start suffering; when they escape danger, we derive enjoyment from their success.

Therefore, watching a movie involves not only enjoyment but also suffering—although we are not aware of the suffering at the time. If the total enjoyment is more than the suffering, we'll be happy, and when the movie ends our body remains well-balanced for a while, however if the total enjoyment is less than the total suffering, we remain out-of-balance for a while. I will use *overbalanced* and *underbalanced* to refer to a net state of happiness or suffering, respectively.

What happens when you're overbalanced or underbalanced? Let's examine this using a funny movie that leaves you happy when the movie ends. This overbalance will counteract some of the sufferings you encounter in the immediate future. For example, you are at home and get bored. You lose your balance, so you make a motion by going to a movie. You enjoy the movie and are consequently happy and overbalanced because your enjoyment surpassed your suffering. A portion of the enjoyment counteracts the suffering you endured before the movie (being bored), while the remainder will counteract some suffering in the near future. For example, you are driving home from the movie theater, and you stop at a red light. Your mind begins to wander (perhaps remembering the movie), and the car behind you honks when the light turns green. Different things could happen according to your level of balance:

If you are extremely happy, you may say, "I'm sorry," whether or not the driver actually can hear you.

If you are happy, you may shrug and wave.

If you are neither happy nor upset, you may say, "Okay, I'm going."

If you are unhappy, you may say, "Give me a second!"

If you are extremely unhappy, you may curse or gesture obscenely.

In this scenario, since you're returning home after the funny movie, you are happy and overbalanced in front of the red light. When the driver behind you, honks, you'll naturally lose your balance; however, because you are overbalanced, your balance drops to a lower level.

Here I must point out that watching a movie is not a motion, but going to a movie is. Listening to music is not a motion, but turning on your stereo is. Enjoyment without motion always creates an overbalance; such enjoyment will counteract previous suffering or the immediate future suffering. Also suffering without motion such as hearing bad news always creates an imbalance that will counteract by last enjoyment or with the next immediate enjoyment.

Objection five:
"When I hear a joke, I laugh. I enjoy without suffering."

The enjoyment of such actions can be divided in two portions. The first potion pays of the past suffering and the second portion places you at a higher level of balance and will counteract the next suffering, When you hear a joke, you lose your balance; you take action by laughing to regain your balance. To understand this better, think about what would happen if you prevent yourself from laughing when you heard a joke. You would become uncomfortable; you are preventing yourself from doing something you want to do. (We have all had this experience in a classroom when someone says something funny, and we try to stop ourselves from laughing; we become uncomfortable and ultimately burst out laughing.) No matter what, we always suffer when we prevent ourselves from doing things we want to do. On the other hand, we suffer from not receiving what we want. We may want to laugh, but we don't allow ourselves the pleasure. It's obvious that we suffer from not receiving the things that we want.

Objection six:

"I know many kids who have comfortable lives and never suffer."

This is a broad statement. Generally speaking, the majority of these kids will suffer from a difficult life for a while at some point in the future if they haven't suffered from studying; those who have suffered from studying hard and are disciplined will enjoy an easier life in the future. However, this is not an appropriate answer to this objection. We have to look at every kid individually.

However, the enjoyment these kids experience could be divided into two different categories.

The first category includes enjoyments such as eating, drinking, and sleeping that are followed by related sufferings of hunger, thirst, and tiredness. The second category includes enjoyments such as playing video games, watching TV, or playing with friends that have no suffering in the present. The sufferings that follow such pleasures await the kids in the future, which is why we worry about children who don't get a good education and aren't disciplined. They will surely suffer for a while in the future. I say "for a while" because they will suffer as much as they enjoyed before. Suffering ends when they counteract the earlier enjoyment and then get used to a difficult life. It's obvious that those who have an education suffer less than those who don't— the present suffering that they experience is far outweighed by future enjoyment. At any rate, the best answer for this objection requires analyzing every kid individually because the lifestyle and living conditions of one child differ from the next; specific factors are always related to a particular kid. Besides, the general childhood enjoyments will be counteracted by the suffering of old age when strength is lost.

Conclusion:

There is always enjoyment for each suffering and vice versa on each action.

Sufferings and enjoyments without actions, eventually, counteract each other.

Chapter 4

WHAT WILL HAPPEN TO US IF NOTHING BOTHERS US?

WE REMAIN MOTIONLESS IF NOTHING BOTHERS US.

What would happen to us if, someday in the future, we were able to remove and destroy all the things that bother us throughout our lifetimes?

Viruses and Bacteria:

If we destroyed viruses and bacteria and thus associated diseases, not only would we never suffer from illnesses, but a vast number of industries and services maintaining our health would be eliminated. There would be no more hospitals to build, no more labs, no more drugs, no more pharmacies, no more doctors, no more medical institutions, and no more health insurance companies. Although now we may not be involved with these activities directly, we must work to pay for such materials and services when we need them. Without any illnesses, we would work less and, consequently, engage in fewer activities.

Envy:

Imagine that one day, we invent technology that could remove envy from human nature. Obviously, not only would we not get hurt by jealous people and not enjoy the friendship of non-envious people, but we would also lose some of our activities. Envy generates motivation, which results in ongoing competition among people. Envy creates the desire for better cars, clothes, homes, furniture, jewelry, watches, hairstyles, hair color, shoes, purses, and so on. If we removed envy from human nature, we would eliminate competition and destroy desires, limiting the production of such materials. If we did not compete, the following activities would be eliminated or limited to our absolute basic needs:

0- Automobile:
Design, fabrication, marketing, repairs, used car services, auto insurance, and other related business and services.

I- Clothing, housing, furniture, home decorations, shoes, purses, jewelry, watches, hair, nails, facial, fitness equipment, cosmetic surgeries, and so on.
Clearly, if we removed envy, we would limit or eliminate the activities of thousands of big companies. We would work less because we would not need these products and services.

Aggression and dishonesty:

If we removed aggressiveness and dishonesty from human nature, not only would we not enjoy the friendship of honest and nonaggressive people, but we would also not need the armed forces (including hundreds of related businesses), police departments, security guards, gates, locks, courts, judges, lawyers, and any related businesses and services. Thus, by removing dishonesty and aggression from human nature, we would eliminate a considerable number of activities from our lives. Although we may not be involved directly with these activities, we would work less because we would no longer have to pay for such materials and services.

Cold and Heat:

Imagine what it would be like if the air temperature was always at a fixed and comfortable level. Clearly, not only would we not suffer from cold and heat, but a large class of related activities of air-conditioning and heater manufacturers, air-conditioning contractors, and businesses that make and distribute warm clothes and heavy fabrics would be limited or eliminated. Thus, if the temperature did not bother us, we would work less than we do now.

Hunger:

If there was plenty of food available without cultivation, a very large line of activities would be eliminated from our lives; consequently, there would be less work for us.

If we were not bothered when we saw our plants dying, we would not water them. Not watering is another lost activity. If a messy room did not bother us, we would not organize the room—another lost activity.

There are hundreds of other unwanted problems and discomforts, and it would be overwhelming to name all of them in this book, but you can eliminate them one after another in your mind and see yourself engaging in fewer and fewer activities, to the point that if you were to eliminate all of them, your motions would fall to zero. At this point, you would feel like a dead person. Now, if you were asked whether diseases, germs, viruses, cold, heat, dishonesty, hunger, envy, greed, supremacy, aggression, and the like are bad, how would you respond? Should we make the world free of such problems, or should we embrace them as a life necessity? **What do we miss if we accept nature as is? What do we gain by changing it? Why are we afraid to face life's realities?**

Are the following attributes bad or good?
- Greed
- Supremacy
- Aggression

How would our lives be today if we had removed the above traits from Steve Jobs and Bill Gates' characters when they were twenty years old?

It is so obvious that we would not have such great improvements in computer and communication technologies if we had removed these attributes from Steve Jobs and Bill Gates' characters.

Talents are grown by curiosity and poverty.

Technologies are grown by greed, supremacy, and aggression.

Problems and discomforts are nature's secret tools to bother and motivate us.

Ultimate comfort inactivates us.
Problems, troubles, and discomforts activate us.
We remain motionless if nothing bothers us.

Conclusion:

Problems and discomforts bother us and generate an imbalance in us. We make motions to regain our balance, and we remain motionless for as long as we do not feel an imbalance. We need problems to be active and alive. [F8]

Chapter 5

WHY WE DO WHAT WE DO?

In the first chapter, we reached the conclusion that, when you lose your balance because of a factor affecting you, you make a motion to regain your balance (fact two). This is a simple action when there is only one factor involved in a motion. However, sometimes when you want to act, another factor prevents you. Consider the following examples.

Example one:
The weekend is here, and your family is going on a picnic. You lose your balance, and you *want* to go with them, but you have an exam on Monday.

You will have a bad feeling (imbalance) if you get a low grade on your exam, resulting in suffering in the future (which we define as feeling *fear*—fear of losing balance and suffering in the future). What would your motion be in this scenario? There is the *want* to go and the *fear* of suffering in the future. If your fear is greater than your want, you'll stay home and study; if your *want* is greater than your *fear*, you will go to the picnic.

Example two:
you are presented with a delicious cake at a party,

Want to eat and *fear* of gaining weight

Loss of balance by seeing delicious cake generates *want* to eat, and visualizing gaining weight creates *fear* (imbalance) of not to eat. You perform motion based on the outcome of your want and your fear.

Fact six: The sum of the outcome of our want and fear determines our motion. [F6]

As a matter of fact, our actions and reactions are based on our balance.

In the above example, there are two opposite forces that determine our motion: motivating force (enjoyment or want) and preventing force (fear of suffering), and the outcome of these forces determines our motion. In other words, we perform motions where there is less suffering or more enjoyment.

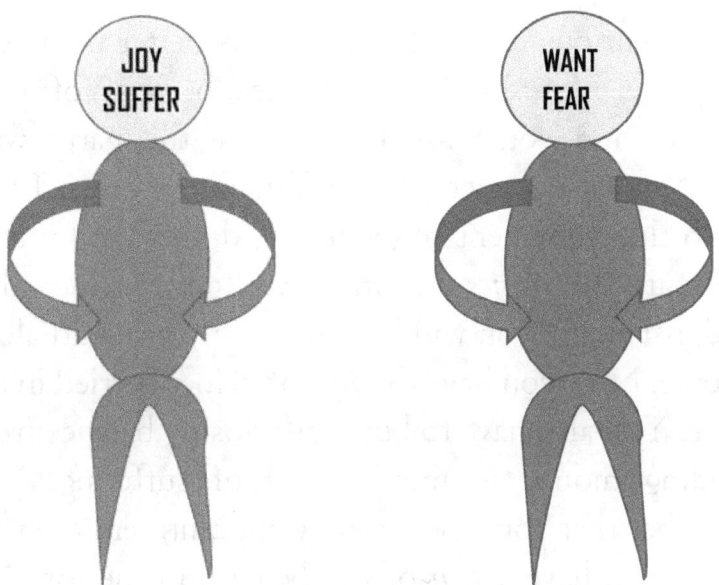

The sum of the outcome of our joy and suffering determines our motion or we can say, the sum of the outcome of want and fear determines our motion

Enjoyment = Motivating force
Suffering = Preventing force

The above formula could be reversed depending on the situation. For example, if you have the potential to gain weight, the fear of suffering from a bad figure motivates you to diet.

In some cases, several wants and fears arise. Consider the following example.

Example three:

There is a party tonight, and you *want* to go. You

also are curious to see if this party is better than your last party. You lose your balance because of your curiosity and you *want* to compare the party with yours. You also *want* to show off your new dress. There is also the enjoyment of gathering, dinner, and music. These are the *wants* that make you go. On the other hand, if you go, you will have to leave your child alone at home. Now you have the *fear* of being worried in the future. You also have to buy a gift; losing balance from spending money is another *fear* of suffering. You remember that you have an early appointment the next day, so you have a *fear* of not being on time for your appointment. You realize that there is a long drive to the party. You have a *fear* of suffering from driving and spending money on gas.

This scenario includes several acting forces (*wants*) and several preventing forces (*fears*). Your mind evaluates the degree of each enjoyment and suffering, in the other words your mind determines the degree of each *want* and *fear* and you follow through with the motion according to the perceived outcomes of these wants and fears. If the total of your wants is greater than the total of your fears, you will go; if the total of your wants is less than the total of your fears, you won't go. **Actually, decisions are nothing but the outcomes of wants and fears.** We will discuss this subject in greater detail in the chapter on How we make decisions.

As the result, our motions are based on our balance. We don't move at all if we don't lose our balance. When we lose our balance, we make a motion to regain it. That's how human nature works.

Fact 7s: The sum of outcome of our wants and fears determines our motions.

Conclusion:

Our actions and reactions are based on our balance. We perform motions only when we lose our balance, and we remain motionless as long as we don't get imbalanced. [F2]

Chapter 6

ROLE OF PERSONALITY, PHYSICALITY MENTALITY AND ENVIRONMENT IN OUR MOTION

Personality characteristics:

Personality is the combination of numerous characteristics, such as envy, sensitivity, honesty, pessimism, responsibility, selfishness, patience, punctuality, generosity, forgetfulness, forgiveness, confidence, laziness, and so on.

When we come into this world, physical characteristics come with us through our genes; one child may have blue eyes, and another has brown eyes. One has light skin and another dark skin. The characteristics of our personalities also come with us when we are born, although environments may affect these characteristics. Weather, nutrition, education, friends, family, relatives, culture, religion and so on could influence our personality characteristics to varying degrees. For instance, if we measure patience from one to ten and a newborn is a seven, the level of his patient may become a five if he grows up in an impatient family or an eight if he grows up in a patient family.

As you read this, you are at an age at which numerous characteristics in your personality have reached a certain degree. These directly affect your motions. For instance, when you see that your friend bought a nice car, your degree of envy will cause you to react in one of several ways. If you are not envious, you will be happy for your friend and say, "It's a nice car. I'm happy for you, and I hope you enjoy the car." If you are a little envious, you may say, "That's nice." If you are rather envious, you may say, "I don't like that type of car." If you are really envious, you may take a screwdriver and scratch the car when nobody is around. If you are extremely envious, you're so imbalanced that only one motion can regain your balance: buying a better car or laughing when your friend's car is destroyed in an accident.

Do you really want to hurt your friend? No. You like your friend, and you enjoy your friendship. The imbalance of envy makes you do something to regain your balance. Where is the option? If we can choose, why do some people choose to hurt their friends? Why can't a jealous person choose to be truly happy and say something nice when a friend buys a nice car?

The degrees of our characteristics has a substantial effect on our wants and fears. The degrees of our characteristics have major role determining the levels of our want or fear. For instance, in the last example, the degree of envy determined the level of want, which determined the motion. Let's analyze this more thoroughly.

When you see your friend has bought a new car, you lose your balance and *want* to say or do something to regain it. If you are not an envious person, you will become really happy, and you *want* to say something nice to regain your balance (if you don't say anything and remain silent, you will feel uncomfortable, and so you must say something nice to regain your balance). If you are a little less of an envious person, you will become a little unhappy and will *want* to say something to regain your balance. Here you may say, "That's nice." If you are an envious person, you will become unhappy, but you still *want* to say something to regain your balance. Here you may say, "I don't like this type of car." If you are really envious, you become really unhappy, and you *want* to say something really nasty or to do something to regain your balance, which is why you may scratch the car.

As you see, the degree of your envy determines the level of your want, which determines your motion, so the level of your want depends on the degree of your envy, which ultimately determines your motion.

Every day we see good friends hurt each other because of their envy. Focusing on their actions, leads us to the fact that the person who hurts his or her friend knows that he or she will suffer, but still hurts the friend. Why? Because suffering from envy made the person so imbalanced that he or she had to do something to restore his or her balance. In this scenario, the *want* to regain balance is greater than the *fear* of suffering seeing the hurt friend.

Let's look at another example to see the effects of our characteristics on our *fears*. The degree of our conservatism determines the level of our fears. For instance, you are at a party and you see a beer that you *want* to drink, but you *fear* getting a ticket as you drive home. The level of your *fear* depends on the degree of your conservatism. If you are not a conservative person, you may drink and drive. If you are a conservative person, you won't drink. If you are somewhat conservative, you may drink a little. As you see, the degree of our personality characteristics determines the level of our fears.

As such, the degree of our personality characteristics plays a big role in determining the level of our *wants* and *fears* as well as in our motions.

Physical characteristics:

The level of our wants and fears also depends on our physical characteristics— height, weight, age, gender, skin color, hair color, strength, cholesterol level, blood sugar, blood pressure, and hundreds of other physical characteristics may affect the level of our wants and fears. For instance, high cholesterol creates a *fear* of eating greasy foods. If you are seventy years old and *want* to move the coffee table in your living room, the *fear* of back pain may stop you from doing so.

Mental conditions:

Mental conditions such as depression, anxiety, obsessive worry, panic, insomnia, manic high, and so on can seriously affect our motions. When we are depressed it's obvious that our motions are totally different than during normal conditions. For example, a woman who normally wants to go shopping may lose her desire to do so when she is depressed.

Environmental conditions:

Environmental factors also affect our wants and fears. For instance, when you get upset at your spouse in the presence of others, you might not argue; but if nobody else is in the room, then you may argue. The presence of others as a condition of the environment affects your motion. To clarify this example, let's conduct an analysis.

Your spouse says something that bothers you, and you lose your balance. You *want* to say something to regain your balance. If there is no other factor preventing you, you may say something to regain your balance. However, the presence of others creates a *fear*—your fear of suffering from people's perceptions or reactions stops you from arguing. Of course, you will argue in the presence of others if the want is greater than the fear.

Clearly, environmental factors (for example, the presence of others) create fears that play a big role in determining our motions. Another example is society's respect for educated people. This respect as a condition of the environment creates a want to become more educated. Environmental factors such as culture, religion, friends, family, geographical locations, weather, and nutrition influence our motions by creating and changing the level of our wants and fears.

Fact seven: The degrees of our personality and physical characteristics, and mental and environmental conditions, determine the levels of our wants and fears on every motion, and the sum of the outcomes of those wants and fears determine our motions. [F7]

The degrees of our characteristics
and environment have major role on our motions

The degree of each personality characteristic depends on the chemical formation of the related gene, just as with physical characteristics. Therefore, personality characteristics are physical conditions. Mental conditions are chemical formations that differ from normal chemical formations. This suggests that physical characteristics, personality characteristics, and mental conditions are all physical which is called DNA.

Now regarding the above consideration, we can reword fact seven:

Conclusion:

The degrees of our characteristics (DNA) and environmental conditions determine the levels of our wants and fears, and the sum of the outcomes of our wants and fears determine our motions. [F7],

Chapter 7

DO WE HAVE FREEDOM OF CHOICE?

A friend of mine was challenging me that we have freedom of choice when she was in my home with some other friends in a small friendly gathering. To prove her claim, she picked up the glass of water and placed it on the right side of the table, and then she picked it up and placed it on the left side of the table and said: "I can place the glass anywhere I want, don't I have the freedom of choice?" I said: "You are one hundred percent right when the location of the glass makes no difference to you, but if you have a slight interest in the location of the glass then that little interest creates a want or a fear in you that causes a specific action. I should say 99% or almost 100% of the time there is an interest or a negative point in our actual daily motions, and that interest or negative point directs us to a specific action that differs from our original choice. The discussion was interrupted at this point because some of the friends wanted to leave. Anyhow, we gathered another day in her home and I observed that she was placing the coffee cups only on the cup mats when she was serving the coffee. I asked her if there was a reason that she placed the cups only on the cup mats not elsewhere on the coffee table. She said the cups may leave spots on the glass top and then I have to clean the glass top of the table. I said, referring to our last unfinished discussion, "So the suffering of cleaning creates a fear, and that's why you place the cups only on the cup mats, not elsewhere on the glass top."

She said how about this: I'm going out tonight, I can choose to wear my blue jacket or wear my red jacket. Analysis: You can choose to wear any of those jackets if there is no preference, but I am almost100% sure that there is some preference here, such as:

o Your boyfriend likes the red on you.

o Tonight is a cold night. You may wear a warmer jacket.

o The jacket you want to wear is wrinkled and there is no time to iron it.

o There is a lost button or a food stain on one of the jackets.

o You wore the red jacket at the last gathering.

o None of these two jackets matches the shirt you like to wear tonight.

I'm sure that in real life similar factors always interfere with your freedom of choice and determine a specific choice for you that differs from your original free choice.

It seems we have freedom of choice, but in reality, there is no choice. Here I would like to get your attention to the following examples of our actual actions to clarify if the freedom of choice exists:

1) How many times did you choose not to pay the car loan payments?

2) How many times did you choose not to pay for your traffic ticket?

3) Thousands of times the salad dressing or food sauce touched your lips; how many times did you choose not to wipe up your lips?

4) How many times did you choose not to wipe off the foggy windshield and continue driving?

5) Thousands of times you were at the traffic light; how many times did you choose not to stop at the red light?

6) How many times did you choose not to pay the house mortgage payments?

7) How many times did you choose not to file your yearly income tax?

8) How many times did you choose not to scratch your forehead when it's itching?

9) How many times did you choose not to say hi-

back when a good friend says hi to you?

10) How many times have you seen a mother choose not to feed her baby when the baby is crying for food?

Focusing on the above actions leads us that there is no freedom of choice. Instead, the potential outcome of the motivating forces and preventing forces determines our motions.

Let's analyze each of the above examples to clarify the fact that freedom of choice doesn't exist:

Example 1

How many times did you choose not to pay the car loan payments?

Fear of getting a low credit score or losing your car makes you pay the car payments, not the freedom of choice.

Example 2

How many times did you choose not to pay for your traffic ticket?

The fear of getting involved with the law authorities makes you pay for the traffic tickets not the freedom of choice.

Example 3

Thousands of times the salad dressing or food sauce touched your lips, how many times did you choose not to wipe up your lips?

You lose your balance and feel uncomfortable when the salad dressing or food sauce touches your lips. You wipe it off to regain your balance. As we discussed before, **we always want to be in balance (comfortable),** and that's our nature. We will feel uncomfortable and imbalanced if we leave the dressing on our lips. Actually, our brain recognizes suffering from imbalance and generates the action of wiping in order to regain balance. As a matter of fact, it is the imbalance of having sauce on your lips that makes you clean your lips not the freedom of choice.

Example 4

How many times did you choose not to wipe off the foggy windshield and continue driving?

The fear of getting into an accident makes you clean the windshield not the freedom of choice.

Example 5

You encounter traffic lights thousands of times. How many times did you choose not to stop at the red light?

Several factors make you stop at the red light: fear of getting a traffic ticket, fear of getting into an accident, and fear of losing self-respect by breaking the law. Your brain evaluates that the total suffering of the three fears is greater than the enjoyment of getting to your destination early. Therefore, you stop at red lights, not by the freedom of choice but because of these three factors. You may pass the red light if your brain evaluates the total fear of three factors is less than not getting to your destination on time. For instance, you receive a call that your child got hurt at home and no one is available to take her to the emergency room. In such a case, your total suffering from the three factors is less than suffering from getting home late, and you pass the red light.

Example 6

How many times did you choose not to pay the house mortgage payments?

Fear of getting a low credit score or losing your house makes you pay the mortgage payments not the freedom of choice.

Example 7

How many times did you choose not to file your yearly income tax?

The fear of getting involved with the tax authorities makes you file your income taxes not the freedom of choice.

Example 8

How many times did you choose not to scratch your forehead when it's itching?

You immediately scratch your itchy forehead to regain your lost balance. Have you ever seen anybody in your entire life choose not to scratch his or her itchy forehead?

Example 9

How many times did you choose not to hi-back when a good friend says hi to you?

The fear of suffering always makes you to hi-back not the freedom of choice.

Example 10

How many times have you seen a mother choose not to feed her baby when the baby is crying for food?

The suffering of the baby's cry creates an imbalance in the mother, and she feeds the baby to regain her balance, so the fear of suffering determines the mother's action, and there is nothing else here to play a role, such as freedom of choice.

Microanalysis:

You said that you believe in freedom of choice, and you can perform actions on any of the above examples by the freedom of choice to prove your claim.

I agree with this experiment, but let's make the experiments on only two of the examples because the results are almost similar. Let's randomly choose example 9 and example 6.

Analysis of example 9; you chose not to hi-back to

a good friend any time when he sees you and says hi to you. It is obvious that you will lose a good friend. Think and find out the actual cause of not saying hi back to your good friend. Is it the freedom of choice or it is the desire to prove that you are right in your belief? We always lose our balance and suffer when someone talks against our beliefs, so the fear of suffering made you not hi back not the freedom of choice.

Analysis of example 6; you chose not to pay the house mortgage payments and you are a few months behind the payments to prove that you have freedom of choice. What happens next? Do you continue not to pay the payments? Is your desire to prove your claim worth losing the house? Or do you make the payments and save the house? Either way, freedom of choice has no role in your action. The outcome of your want to prove your claim and your fear of losing the house determine your motion.

A. You didn't pay the payments and you lost the house. In this case, your want to prove your claim was greater than the fear of losing the house.
B. You made the payments and saved the house. In this case, your want to prove your claim was less than the fear of losing the house.

Conclusion:

Freedom of choice doesn't exist; instead, the potential outcomes of our motivating forces and preventing forces determine our motions. [F9]

Chapter 8

HOW WE MAKE DECISIONS
Free Will Analysis

What is free will? How do you describe free will? Some say it gives us the power to make decisions. Others say it allows us to do whatever we want. Still, others believe free will gives us the option to choose.

I suggest doing an easy experiment to prepare your mind for the free will discussion. Stop brushing your teeth if you believe in free will. I'm sure you can decide to avoid brushing for a few nights or a few weeks to prove that you have the power of the decision-making, but eventually, the fear of having bad smelling mouth and the fear of losing your teeth will become greater than the desire to prove your claim, and you will start brushing your teeth again.

Before I talk about free will, let's complete the discussion from chapter I.

I said that when a factor affects us, we lose our balance and make a motion to regain our balance (fact two). This is what normally happens, but sometimes other factors are involved.

Example one:

You are at home and become hungry. There is no food at home and its cold outside. The hunger factor makes you lose your balance, which creates a *want* in you. You want to go buy some food, but there is also another factor—the cold, which will make you lose your balance if you go outside. This *fear*—the fear of suffering from cold also plays a role. One factor is the *want* (motivating force) that makes you go to market, and the other factor is the *fear* (Preventing force), which prevents you from going to the market. Now, what happens? If you are only a little hungry, you won't go to buy food; however, eventually, your hunger will increase to the point that you are compelled to go out to buy some food. As a matter of fact, you will be driven by whichever factor is greater. When the *fear* is greater than the want, you won't carry out the motion. But, as the *want* becomes greater than the fear, you will carry out the motion. In other words, there are two sufferings: suffering from hunger and suffering from cold. Your mind recognizes exactly which factor is greater, and you carry out the motion where there is less suffering.

Example two:

You have high cholesterol, but you have been presented with some delicious, high-cholesterol food. How would you act? Would you eat it or not?

If you are a little hungry, you may not eat it or eat a little, but you will eat the whole dish if you are extremely hungry. Where is free will in such a situation? How do you analyze your freedom of choice? The only factors involved are the *want* to eat (enjoyment) and the *fear* of harming your body (suffering). Unknowingly, your brain recognizes the degree of enjoyment and suffering: enjoyment from eating and fear of suffering from hurting your health. Your action will be according to the outcome of the enjoyment and suffering. The mind's evaluation is constant and immediate. Our experiences and information play a significant role in the result of the evaluation. If you know that high cholesterol could kill you, your *fear* will be much greater than if you lacked knowledge about it. Thus, your brain creates different degrees of fear based on the available information. Still, your motion is based on the sum of the outcome of your wants and fears, we also can say, we perform actions where there is more enjoyment or less suffering.

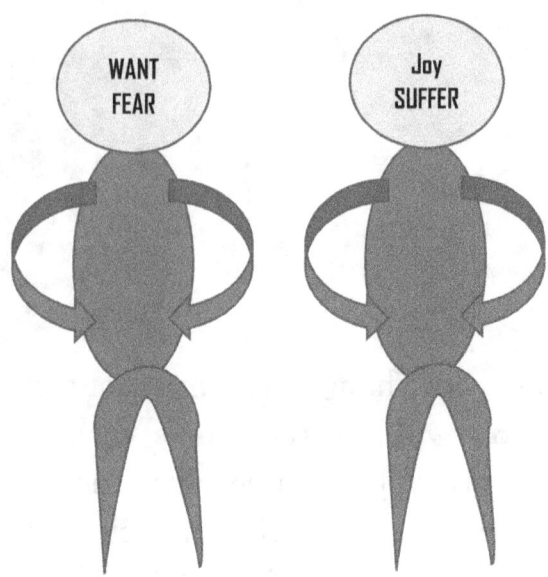

We make motions according to the outcome of our joy and suffering sums or we can say, we make motion according to the outcome of our want and fear sum

Example three:

Your child is playing video games even though he has an exam tomorrow. You *want* him to pass the exam, but there is a *fear* that if you stop him from playing, you will suffer from his state of upset. What do you do? Your mind automatically calculates which is greater: the suffering from seeing him fail the exam or the suffering from his state of upset. Your body will carry out the action that involves less suffering. We have seen parents' actions in such situations. Some stop their children from playing because they feel the fear of future suffering if their children have difficult lives.

Others act differently. The *want* of seeing their children happy during the present time is greater than *fear* of suffering in the future. As a result, they don't stop their children from playing. In addition, some parents may ask their children to stop, but the children start crying, and the parents say it's okay to play. Here, the present suffering from the child's upset becomes greater than the visualized suffering of the future of the unsuccessful child.

Some argue that parents act either logically or emotionally. I believe it's not a matter of logic or emotion, but the potential outcome of *want* (seeing your child happy during the present time) and *fear* (visualizing the suffering in the future), whichever is greater determine the motion. The potential outcome varies according to the characteristics of our personalities, which I discussed in detail in the chapter on the role of personality and environment in our motions.

Each of these examples involved only two factors. Sometimes more than two factors are involved in a motion.

Example four:

You see a car at a dealership, and you strongly lose your balance because it's the car you want. You want to buy the car, but you have fears, such as high payments, high maintenance, repair costs, and high gas consumption. Your mind recognizes the degree of each fear, and you perform motion according to the weighing of your want against the four fears. If the suffering from not having the car is greater than the sum of the four fears, you will buy the car. If the total sufferings of four fears is greater than the desire to have the car, you will not buy the car.

Sometimes additional factors make the *want* greater; for example, if you buy the car, you'll please your girlfriend or your friends and relatives will look up to you. You may have the feeling that you want to prove to yourself that you can have a car like that. As a result, you may act differently because there are more wants.

At this point, I will provide you with several examples. Analyze where free will is. What is decision-making? How do we perform motions?

Example five:

You are at home, and you're tired and want to rest. Your friend calls and asks you to go for a walk. You explain that you are tired and *want* to stay home and rest. Your friend then says that John is coming too, and today is his last day in town. You get up to go for the walk because John is your good friend, and you *want* to see him before he leaves.

Before you head out the door, the telephone rings again. It's your mother calling from the airport. She tells you that your brother couldn't pick her up, so she needs a ride home. You agree to pick her up in thirty minutes.

As you are leaving, the telephone rings again; this time, it's your teenage daughter. She tells you that she was just in a car accident. She is okay, but she can't drive the car. You tell her to stay put and you'll be there in a few minutes.

Before you can get out the door, you hear your three-year-old son screaming and crying. He has fallen down the stairs and seems to have broken his arm. You have to take him to the hospital.

In this example, you are faced with five decisions to make or five options to choose from. What will you do? Your *want* is to stay home and relax. Is that free will? You may stay home and relax if your brain doesn't function properly. However, you will not stay home, go to see your friends, give a ride to your mother, or go to help your teenage daughter. Instead, you will take your three-year-old son to the hospital. Why? Because the *want* of seeing your son with a healed arm is greater than any of the other wants. We can say that the suffering from seeing the broken arm is greater than the fear of suffering from seeing the unhappy daughter or upset mother. There is no such a thing as an independent power called free will; there is only the recognition and evaluation of wants and fears in our minds—a continuous and fast logical evaluation of joy and suffering (wants and fears). We always act in relation to where there is less suffering or more enjoyment.

The best examples are those that occur in your daily life. Look at your life today. You have performed hundreds of motions already. Most involved one factor, but some involved two or more factors to which you reacted. Analyze your motions and see if you could act differently.

Example six:

There are events in our life that we remember in vivid detail, such as when we married, bought a house, chose a field of study, bought a car, or moved to another city. Choose one of these events, making sure you remember the details, the conditions and factors involved at the time of the action, and what led you to take that action. Now imagine that you could go back to that time and all the factors and conditions were exactly the same. You are the same person and your experience level and knowledge are exactly as they were. What would your action be this time?

Every single person whom I've ever asked this

question has responded, "I would do the same."

Let's analyze a common event: buying a car. Some of the factors involved in this action include your budget for a down payment, your income, your credit, your other payments, the type of car you like, gas mileage, the color you like, the shape of the car you like, and your driving distance to work. You checked out several cars and evaluated all the factors before finally purchasing that particular car. If you go back to that time and all the factors remain the same and your knowledge and experience are also the same, there is no reason that you would act differently. If you were to go back a hundred times with the exact same conditions, you would buy that same car again and again.

Now consider going back to that time and one or more factors were not the same. For example, now you have the knowledge that the resale value of the car is very low, the parts and repairs are expensive, you have a lower income now, or you need a larger down payment. The car you would buy will not be the same. Now you see that if the condition differs, you will buy a different car—so you didn't have an option to choose. The condition and factors at that precise moment made you buy that car. You think that you *decided* to buy that car, but the reality is that there is no such decision-making power. The fact is, that you wished to buy the latest model and a fancy car, but the condition and factors placed you in a situation in which you couldn't buy your dream car. The resultant of your wants and fears made you buy that car.

Example seven:

Recall the details surrounding the time when you got married and imagine going back to that time. All the conditions are the same—your experience, knowledge, and age. What would you do this time? Do you marry the same person or not? If not, why?

If your answer is no, you have a reason. If you analyze that reason, you will see that the reason is not one of the same conditions present when you got married and it's a different condition. You need to place yourself in the exact same conditions, including your personality, physicality, mental and environmental conditions, in order to find the true answer.

Example eight:

The previous two examples were first-degree and outstanding actions. Let's remember some of the second-degree actions such as going on a vacation or breaking up with a friend. Place yourself in that time with all the same conditions and see if you would act differently.

Let's remember a third-degree decision from the last

few days or the last few hours. This would be something simple like eating in a restaurant, going to a movie, taking your medication, visiting a friend, or wiping your car's windshield. We all think that we have decision-making power and have the option to choose, but the reality is different from our beliefs. For instance, remember the last time you wiped your foggy car windshield. If you go back to the same time in the same exact condition, would you continue driving with a foggy windshield? Obviously not, and you act exactly as you did before; as a matter of fact, the foggy windshield creates the fear of an accident and makes you lose your balance, and you wipe the windshield to regain your balance.

We act precisely according to our specifications and properties (DNA: personality characteristics and physical characteristics and mental conditions) affected by the environment.

Imagine that you can travel through time to five days in the past and assume that all conditions remain the same. What would your actions be this time? Would you act differently or you act one after another exactly as what append before?

Now, what would your answers be to the following questions?

(1) You are a very envious person, and your friend buys a very nice house. Could you choose to be happy from the bottom of your heart and say something sincerely nice to your friend?

(2) You are a pessimistic person, and you see your girlfriend with another man. Could you think that there is nothing going on and not bother her with questions about what you saw?

(3) Your child is crying for food. Could you ignore the cries and chose not feed her?

(4) Could Mozart choose to become Al Capone?

(5) Could Al Capone choose to become Mozart?

(6) Could Gandhi choose to become Hitler?

(7) Could Abraham Lincoln choose to become the American singer, Michael Jackson?

(8) Can you choose to become a supermodel if you don't have the required looks and figures?

I am sure that your answers are no. You may object and say that these are extraordinary examples; I believe you are right. Let's try some simple examples.

You arrive at a party where vodka is being served. You lose your balance and *want* to drink it, but then you see whiskey on the other side of the counter. You lose your balance a little more because whiskey gives you more pleasure than vodka. Which one would you take to drink? Of course, you would drink whiskey if there is no other preventing factor involved. So, your action depends on where there is more pleasure.

We perform motions where there is more pleasure; in other words, we perform motions where there is more want.

If this scene happens another hundred times with the same conditions, you would always choose whiskey because the *want* for whiskey is greater than the *want* for vodka.

However, one day you drink vodka and discover another factor that would make you drink vodka. For example, there may not be enough whiskey for everyone, or this time you have a greater *want* to try vodka, or perhaps your doctor told you to lower your consumption of whiskey. As you see, you follow the result of your brain evaluation. Your mind recognizes that there is more pleasure in whiskey, so you choose it *if there is no other preventing factor involved.* However, when you fear embarrassment that if you drink the whiskey and there will be none left for others, you choose to have the vodka.

Now, before you drink the whiskey you remember

that you have to drive. Here is the fear of getting a ticket (an environmental condition). What would you do? Different things may happen depending on your personality.

If you are a careless or risky person, you may drink and later drive because your *want* to drink is greater than your *fear* of getting a ticket. If you are a caring or conservative person, you wouldn't drink at all because the *fear* of getting a ticket is greater than the *want* to drink. If you are in between these two, you may drink a little. Here you see that the characteristics of our personalities affect our motions. Here your want is to drink and have fun, but characteristics such as caring and disciplined behavior prevent you from doing the things that you want to do, so when the conditions change our actions, no option exists. We act according to the potential outcomes of our wants and fears sums not free will.

Considering the previous analysis, four groups of factors determine the level of wants and fears.

I-Personality characteristics:
Confidence, aggression, pessimism, optimism, creativity, forgiveness, sensitivity, reliability, flexibility, envy, calmness, tolerance, selfishness, patience, and so on.

2-Physical Characteristics:

Height, weight, age blood pressure, blood sugar level, cholesterol level, back pain, skin color, gender, strength, etc.

3-Mental Conditions:

Depression, anxiety, inattention, insomnia, and other mental conditions.

4-Environment:

Friends, relatives, society, assets, geographic location, education, experience, TV programs, pollution density, political conditions, and so on.

The degree of each personality characteristic depends on the chemical formation of the related gene, just like physical characteristics; therefore, personality characteristics are all physical conditions. Mental conditions involve a different chemical formation than normal. Yet these mental conditions are physical as well. Thus, the word *characteristics* in this book refer to all personality and physical characteristics, as well as mental conditions.

Fact 7-s: The degrees of our characteristics (DNA) and the environmental conditions determine the levels of our wants and fears, and the sum of the outcomes of those wants and fears determine our motions.

There is no independent power called free will

beyond human bodies.

Here I would like to give you some common examples that most of us have experienced:

A- You see a girl in a nightclub, she seems nice and pretty, and you *want* to talk to her and date her, but you can't approach her because you *fear* rejection. You drink a couple of beers and then easily start a conversation with the girl. What happened here? If you have free will to open conversations with people, why couldn't you? Did the beers change free will? And if so, how do you describe free will, that could change by drinking beer? The reality is that alcohol lowered the fear of rejection. The *fear* of rejection was greater than the *want* of dating before drinking the beers, that's why you couldn't approach the girl, but the alcohol made your *fear* less than the *want* and that's why you made the motion.

I would like to know if you see any other factors involved in this transaction besides want and fear? All my life, I have conducted experiments and research about human motions and I have found nothing but *wants* and *fears.*

B- You haven't taken your car to the carwash for a long time; every day you say to yourself, "I'll do it tomorrow." The *fear* of losing comfort (laziness) and/or the *fear* of spending the money or time prevented you from taking the car to the carwash. You then receive a call from your favorite girl and set your first date for tonight. You take the car to the carwash before you meet with your date. There were many days you wanted to wash your car. Why didn't free will work here? Because there is no free will. The fact is that the *fear* of losing comfort and/or the *fear* of spending money and time was greater than the *want* of a clean car. You made the motion when the want became greater than the fear.

C- Establishing a long-term relationship is always a process of getting to know our partner. I often hear people tell each other, "I'd like to get to know him." We always spend time to become well acquainted with the personality of our future partner. We want to find the degree of each personality characteristic—the degree of pessimism, jealousy, selfishness, forgiveness, kindness, tolerance, responsibility, honesty, sensitivity, stinginess, and so on; why do we want that? Because we believe that people act and react according to the degree of their characteristics. If we believe that people act by free will, why are we worried about their bad traits? We must believe that a jealous person can act like a non-jealous person or a pessimist can act as an optimist or an impatient person can choose to be patient or a selfish person to act generously. Don't we believe that humans act with no self-control? If we change the genes of a person's characteristics, will he act differently or not? Here I have two simple questions:

If we change the gene of a short-tempered man, does he boil again?

If we change the broken water pump of a car, does it boil again?

Our genes are like our built-in components and our motions depend on the properties of our genes (characteristics).

Education, experience, and other environmental factors could change our characteristics. Obviously, we would act differently with the changed characteristics, but still, we would perform motions according to the outcomes of the wants and fears sums with the changed characteristics. The degree of changed characteristics determines a different level of wants and fears and again we perform motions according to the sum of the outcomes of our wants and fears. For instance, you easily lend money to others if you have a reasonable degree of trust, but when someone does not return the borrowed money, then your degree of trust decreases, and a higher level of *fear* develops when you *want* to lend money to the next person. So, obviously, your motion could be different than before, but still, you'll act accordingly to the sum of the outcome of the want and fear.

Our brains constantly evaluate the levels of the enjoyment and suffering, and we perform motions when there is more enjoyment or less suffering.

Let's conduct an experiment together to clarify the existence of the free will. In this experiment, you will make a statement and ask a question, then I will respond, and you will either make a motion or you won't. In this experiment, you are the one who will begin to challenge me to prove that I'm wrong. You must imagine that I'm sitting in front of you, and it's very important that you act when it's time to act. Now let's start.

Your statement and question: "Eddie, you are saying that there is no free will. I can stand up now if I choose to do so. Can I, or can't I?"

My answer: "No, you can't do that."

Your motion

Here you performed a motion or remained immobile.

In response to this, one of the following happened:

A. You stood up.

or

B. You didn't stand up.

Before we analyze these two cases, I recommend that you first read the one that relates to your motion.

A. You stood up.

If you really want to know the free will exists or not, it's important that you honestly evaluate what made you stand up. I suggest that you stop reading my analysis and think deeply to determine those factors yourself.

The analysis of case one is very simple. All our lives we have learned to believe strongly in free will. We become defensive when someone talks against our beliefs. We lose our balance if anybody says anything against our beliefs, and we *want* to prove that we are right.

It went against your beliefs when I told you that you cannot stand up; you lost your balance, so it generated a *want* in you. You wanted to prove that you were right, so you stood up. When you lost your balance, you suffered; when you stood up, you enjoyed (you said, "You see, I'm right") and regained your balance.

Thus, free will does not exist. There is no option or

decision-making power. You simply lost your balance and acted to regain your balance.

B. You didn't stand up.

In this case, one of the following happened:

(1) You weren't interested in this experiment.

(2) You got mad at this experiment and stopped reading.

(3) You got mad at this experiment and continued reading.

(4) You guessed what I would say and continued reading to see the results.

(5) You believed that free will does not exist and continued to read in order to compare my analysis with yours.

(6) You assumed a trick was involved and acted conservatively by not standing up.

(7) You were in a location (for example, a car) in which you couldn't stand up.

(8) You are disabled or ill and couldn't stand up.

(9) Another factor interrupted you, and you

couldn't complete the experiment.

I'm going to analyze this one by one to identify the factors in each. You will have a better understanding if you first read the analysis that relates to your particular action (or lack thereof).

(1) You weren't interested in this experiment. There was no factor to make you lose your balance. Obviously, when you don't lose your balance, you perform no motion.

(2) You got mad at this experiment and stopped reading. When you got mad, you lost your balance; you stopped reading and perhaps called me names to regain your balance or stopped reading to prevent yourself from becoming even angrier (either way, you lost your balance and acted to regain your balance).

(3) You got mad at this experiment and continued reading. You became curious to see the results of this experiment. Your curiosity made you lose your balance, and you acted (continued reading) to regain it.

(4) You guessed what I'd say and continued reading to see the results. You became curious (as a personality factor) to see the results. Your curiosity made you lose your balance, and you acted (read for the results) to

regain it.

(5) You believed that free will does not exist and continued reading in order to compare my analysis with yours. You became curious to see my analysis. Your curiosity made you lose your balance, so you acted (continued reading) to regain it.

(6) You assumed a trick was involved and acted conservatively (as a personality factor) by not standing up. You didn't want to make a fool of yourself; the fear of feeling bad (making a fool of yourself) made you act conservatively. The fear determined your motion.

(7) You were in a location (for example, a car) in which you couldn't stand up. The condition made it impossible for you to stand up. The condition of the environment determined your motion.

(8) You are disabled or ill and couldn't stand up. Your physical condition determined your motion.

(9) Another factor interrupted you, and you couldn't complete the experiment. The telephone may have rung, you may have become hungry, or you may have had an appointment. Whatever factor stopped you from doing the experiment it had a greater effect on you than the experiment itself. For example, if your phone rang and you answered it, your fear of suffering from missing the call was greater than your want to

continue the experiment. Unknowingly, it came to your mind that you could do the experiment later, but you didn't want to miss a call that may be important.

Experiment Two:

Let's repeat this experiment. Let's travel in time and go back to the moment when you were reading this experiment in the chapter on Free Will, assuming that all conditions remain the same.

You say, "Eddie, you are saying that there is no free will. I can stand up now if I choose to do so. Can I or can't I"? What would your motion be this time when I told you that no, you can't do that. Remember that all conditions remain the same, and you don't have any knowledge of the results. Would you act differently or you do exactly as you did the first time? I'm sure your answer is the same. Thus, there is no free will or option in your motion.

Let's remember the things that you were doing a few hours ago: anything, such as taking a shower, putting on makeup, shaving, going to the bank, visiting a friend, arguing with your employer or employee, laughing, crying, or locking your car. Place yourself back in that time and imagine that all the conditions for any of these actions remained the same. You would repeat each of motions exactly as they occurred before and return to this precise moment with all the same exact conditions surrounding you as they do now. It would be the same for your motions yesterday, last month, last year, and all the way back to when you entered this world. So, if you were to go back to the time that you were born and if all conditions remained the same, you would perform the same motions one after another up to this moment, So, *the sum of the outcomes of our wants and fears determine our motions not free will.*

Fact seven: The degrees of our personality and physical characteristics, as well as mental and environmental conditions, determine the level of our wants and fears and the sum of the outcomes of those wants and fears determine our motions.

Conclusion:

   There is no independent power called free will beyond human bodies. [F10]

Chapter 9

EXPERIMENTS OF NO FREE WILL

I insist that you practice some of the following experiments to see if free will exists.

1. Quit eating bread or rice as long as you live (rice for Asian people).
2. You may say this is very difficult, and I agree with you. Quit eating sweets for the rest of your life.
3. You may say eating food is a biological need. Let's practice something easier that would not be considered biological. How about quitting flushing your toilet all the time if you live by yourself.
4. The previous experiment is very difficult too. Let's do this very easy experiment: Avoid cleaning your bathroom mirror and other mirrors at home for good (you must live by yourself to do this experiment). Now imagine the feeling of trying to put on your morning make-up after ten years of using a mirror that has never been cleaned.
5. Decide not to lock your front door forever. This experiment is not for residents of secured homes or homes in secured areas.
6. Prevent yourself from sympathizing and helping an injured child when you see one in an accident.
7. Leave your car always unlocked wherever you park

while leaving your usual valuables in the car.

8. Decide not to wash or change your underwear for as long as you live.

9. Stop brushing your teeth for good.

10. Quit saying hi back when your friends say hi to you.

11. Never stop at the stop signs when you drive.

12. Quit washing your hair as long as you live.

13. Try not to feed your baby when she or he cries for food.

14. Never raise your voice as long as you live, regardless of how angry you become.

For the accurate results of the above experiments, you must disregard Rafii's challenging words. Challenging your beliefs would generate additional want or a fear interfering with the outcomes of your experimentation. When Rafii says something against your belief, he creates a want in you, forcing you to prove that he is wrong. This want interferes with the expected outcome of your actions or reactions; therefore, any of the above experiments must be done disregarding Rafii's challenging words. For example, if you decide to quit brushing your teeth regardless of Rafii's challenging words, you may start brushing your teeth in a week, but if you quit brushing your teeth because of Rafii's challenge, it may take several weeks to start brushing again.

If you succeed in some of the following experiments, you will be one step forward in proving the existence of free will, remember that each practice is for a lifetime.

Free will experiments on personality characteristics:

Quit being stingy if you have a stingy personality.

Decide to be composed when your lover cheats on you.

Decide to be an optimist if you are a pessimist.

Decide to become dynamic if you are a slow person and vice versa.

Decide not to be picky if you have a picky personality.

Decide to become flexible if you have a rigid personality.

Decide not to be stubborn if you have a stubborn personality.

Try not to be suspicious if you have a suspicious personality.

Quit being perfect on the things you do, if you are a perfectionist.

Decide to change your nature and enjoy sour foods if you hate sour foods.

Decide to change your nature and wear red clothes if you hate red.

Decide to change your nature and become a patient person if you have an impatient personality.

Change your nature and try to become a late person if you are an early person.

Change your nature and try to become a calm person if you have an aggressive personality.

Avoid noticing details if you have a detailed-oriented personality.

Decide to be dishonest at all times if you have an honest personality.

Decide to become irresponsible if you have a responsible personality.

Decide to become a high self-steam person if you lack a self-steam personality.

Decide to become non-sensitive if you have a sensitive personality.

Decide to be tolerant if you have a non-tolerant personality.

Use your free will and decide to enjoy the mother-in-law's company while you hate her.

Try to be reliable if you have a lousy personality.

Decide to act normal when you have a phobia.

Decide to become easy to please if you have a picky personality.

Avoid going into details if you are detailed-oriented.

Quit being jealous if you are jealous of others.

Try to be calm in the arguments if you're short-tempered and never get upset regardless of how others make you angry.

Quit being generous if you have a generous personality and choose to be stingy.

Avoid the enjoyment of gossiping, and stop gossiping forever if you are a gossiper.

Decide to be quiet at all times if you have a talkative personality.

Decide to be slow if you have a rush personality.

Try to switch your easy-going personality with an active personality and become an active person if you have an easy-going personality.

Decide, always to act normal if you are a conservator.

Free will experiments on physical characteristics:

Decide to become a delayed ejaculator if you are a premature ejaculator by your nature.

Decide to memorize everything if you have Alzheimer's.

Decide to act normal if you are a bipolar person.

Try to become a world champion runner if you have short legs.

Try to compete with Einstein if you don't have the brain.

Decide to become a supermodel if you are short and don't have the required look.

Always accept jobs that require heavy lifting if you have chronic back pain.

Decide to become a worldwide singer if you don't have the voice.

Decide to be fat if you have skinny nature.

Try to compose like Mozart if you don't have the talent.

Decide to switch your sex drive from low to high if you naturally have a low sex drive.

Free will experiments on mental conditions:

Stop being worried when you have anxiety.

Decide to sleep calmly if you have restless legs.

Try to enjoy life and become cheerful when you are in the depression phase.

Decide to sleep well when you have insomnia.

Decide to become rational when you have a phobia.

Free will experiments on the environment:

Decide to be happy and cheerful when a beloved one passes away.

Decide to be happy and cheerful when you are bankrupt.

Decide not to wear shoes as long as you live.

Decide to dress sexy in a fundamental religious government country.

Never use a heater or fire to warm up your home when you live in a cold climate.

Decide to repair your broken mobile phone when you know nothing about mobile phone technology.

When trapped in a traffic jam, decide to get to your destination on time.

Decide to be calm, relaxed, and cheerful when the driver drives madly.

Stop using a car or bus. Walk twenty miles to the invited party while your family goes in a car.

Quit taking your dog outside to pee and poo and let it pee and poo inside your home.

Decide to be on time for your appointment when your car breaks down on the road.

Try to live like Bill Gates while earning a low income as a cashier.

Never buy anything from a grocery store and try to produce your needs by yourself.

Stop using the mobile phone for good.

You may practice changing your unwanted actions and reactions, and although you can change them to a certain degree, still your actions and reactions depend on the outcomes of your wants and fears. For instance, if you are short-tempered and call your wife bad names in the presence of others when an argument occurs, you may try not to call her bad names by increasing your fear of being perceived negatively in the presence of others. However, this is not the control of your short temper; rather, the fear of people's perception becomes greater than your want to call your wife bad names.

You may also take some medication to control your mental conditions, but this does not help you develop your control of emotions, as the medication only normalizes your unwanted actions and reactions.

Conclusion:

We all have certain degrees of each personality, physicality and mentality characteristics, that determine the level of our wants and fears affected by the environment conditions. As the result, sum of the outcomes of those wants and fears determine our motions, not free will. In other words:

Our DNA and the environment conditions determine our motions, not free will.

There is no independent power called free will beyond our bodies.

Chapter 10

DO WE HAVE SELF CONTROL?

What is self-control? Most people believe that there is an independent power in us that allows us to control our emotions and desires. Does such power exist in human nature? Here I would like to bring to your attention the following examples:

Example one:

Allen was arguing that he quit drinking alcohol. He said that he had been an alcoholic for years, and one day, he decided to quit drinking. Isn't this self-control? It seems that he had power over his desire (want). What is this power? Where is this power? And why can't we use it at all times?

He explained that his wife was always unhappy with his drinking habit. He tried to quit a few times, but he started drinking again. Finally, when his wife decided to divorce him, then he quit drinking for good. The analysis of Allen's self-control is simple, *want* to drink and *fear* upsetting his wife. In the beginning, the fear was less than the want, so he continued drinking. but when the *fear* of losing his wife and family became greater than the *want,* he then quit drinking. Do you see any other factors involved in these motions besides want and fear? There is no factor such as self-control.

Simply the sum of the outcome of *want* and *fear* determined the motion.

Example two:

Pierre decided to quit smoking twenty years ago. He said he tried to quit smoking many times when he was young, but he didn't succeed. One day when he was holding his one-year-old daughter in his arms, the cigarette accidentally touched the daughter's face and burned her cheek. Then he decided to quit smoking for good. The question is: if there is such power as decision-making or self-control, why didn't Pierre succeed in quitting smoking in the past? It's obvious that there are no other factors besides *want* and *fear* in these motions. He failed to quit in the beginning because the *fear* of losing his health and/or *fear* of spending money on cigarettes was less than the *want* to smoke. The cry of the innocent child gave him such guilt, however, that he couldn't forgive himself. He quit smoking for good because the *fear* of suffering by remembering his negligence was always greater than the *want* to smoke.

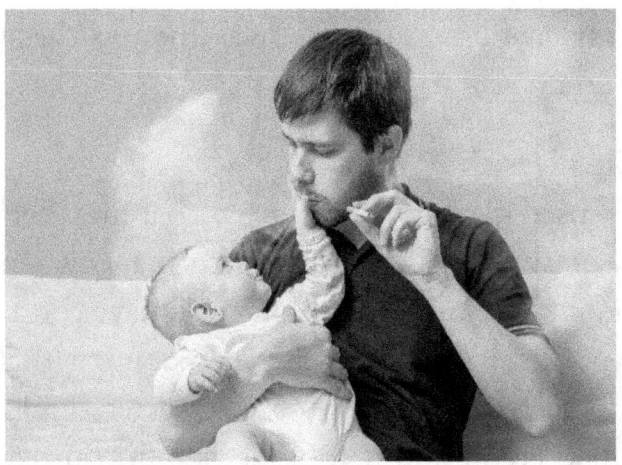

The sum of the outcomes of our wants and fears,
determine our motions, not the power of self-control.

Example three:

You burp when you are by yourself, but you don't burp in the presence of others. You think that you have the power to control your desire (*want* to burp), but in reality, there is no such power. The *fear* of suffering from people's perceptions or reactions is greater than the *want* to burp. That's why you don't burp when others are present.

Example four: You always argue with your husband at home, but you control yourself with him when the two of you are at a party. There is no self-control here. It's the *fear* of people's perception that is greater than the *want* to argue, that's why you don't argue in public.

You argue at home easily because there is the *want* to argue and no *fear* to prevent you.

Example Five:

Lisa used to buy clothes by charging her credit cards. She was always behind with her payments and her mom kept telling her, "Don't you have control over your emotions?" Finally, Lisa ended up with lots of debt and filed for bankruptcy. She suffered a lot from having bad credit and it took her about fourteen years to rebuild it. It seems as though she would have control over her emotions after this long period of suffering, yes? but, actually, there is no power to control her emotions. It's the *fear* of losing her good credit that is greater than the *want* of having beautiful clothes.

Any other self-control example you might have in mind is the result of *want* and *fear*. In other words, we make motions where there is less suffering or more enjoyment.

Conclusion:

There is no self-control or power of decision-making in human nature, instead the sum of the outcomes of our wants and fears determine our motions. [F11]

Chapter 11

WILLPOWER

I would say willpower exists in the meaning of willpower itself. It is the power of want or desire. The power of will is not a fixed sum of power in a person. The degree of willpower is the same as the level of want. This varies from person to person and depends on the type of desire. For instance, the level of want to have a sports car varies from one teenager to another, and the level of want in a teenager to have a sports car is higher than the level of his want to have a house. When the teenager becomes older, his willpower to have a sports car becomes less than his willpower to have a house.

As a matter of fact, when a factor affects us, we lose our balance, and the level of our imbalance is the level of our want. The level of imbalance depends on four groups of factors:

1- Degrees of personality characteristics.
2- Degrees of Physical characteristics.
3- Degrees of Mental conditions.
4- Environmental conditions.

We need to have the willpower to achieve a goal or desire. To achieve our desires, willpower must be greater than the preventing forces (fears). For instance, a woman who has a certain level of willpower toward over shopping (fear of overspending) will experience various reactions when she sees a beautiful dress in a shopping mall. If her level of *want* for the dress is less than her willpower (*fear* of overspending) she won't buy the dress, but if the *want* is greater than the willpower then she will buy the dress.

The willpower toward over shopping is increased by experiencing the difficulties of payments. People who have experienced debt, bad credit, and payment difficulties may have more willpower to resist over shopping than before their debt experience. So, willpower is not a fixed sum of power in a person and can vary based on the effects of environmental factors such as education, experience, culture, and so on.

Another example is the willpower to lose weight. The willpower to lose weight is not a fixed sum of power and varies according to effects from the environment. A girl with the willpower to exercise and diet before marriage may find her willpower eroding after marriage. As a matter of fact, the factors involved are the *want* to be in shape and the *fear* of discomfort from diet and exercise. She performs motions according to the outcome of her *want* and *fear* sum.

Another subject of interest is encouragement. Encouraging is nothing but increasing and decreasing want or fear in people. For example, when we talk to our kids and encourage them to study, we increase their want to have a better future or we increase their fear of a difficult life in the future.

Jenny told her boyfriend Jack, "If you don't come home on time, I'll leave you." Jenny increased fear in Jack to encourage him to be on time.

My doctor told me that fish oil is good for my health. He encouraged me to take fish oil capsules by increasing the want in me for a healthier body.

The threat is another form of encouragement, usually negative encouragement. By treating, we increase *fear* of punishment to encourage others toward action or to prevent an action.

Conclusion:

Willpower is a necessary force for achieving a goal, but it is not the sole factor, the sum of the outcomes of our willpower and preventing forces determine our motions. [F12]

Chapter 12

PERSONALITY CHARACTERISTICS

Our personalities are formed from numerous characteristics, and we have a certain degree of each characteristic. Evaluating characteristics as positive or negative traits is not a proper way to determine one's qualifications. For example, we judge people as dishonest or honest. A dishonest person cannot be trusted at all (zero trust), and an honest person can be trusted one hundred percent. Dividing people into only two groups creates an incorrect evaluation. However, the degree of trust could vary from 0 to 100, but each person has a certain degree of trust. There is no tool available to measure the degree of a characteristic to determine the property of a person, but we can evaluate roughly the degree of each character in a person from 0 to 100. The following is a list of some characteristics.

A careful study of the traits leads us to the importance of the role of the characteristics in performing motions.

Personality Characteristics
Short List

Characteristics	Opposites
accepts authority, loyal, devoted	rebellious
accepts what's given	ignores, rejects what's given
adventurous	conventional
affectionate	distant, cold, aloof
alert	dull
ambitious	non-ambitious
candid	closed, guarded, secretive
calm	excitable, nervous
caring	uncaring, callous
change accept	reject change
cheerful	cheerless, gloomy, sour, grumpy

clean	dirty, unkempt
clear thoughts	muddled thoughts, confused
completes	leaves hanging, doesn't complete
comprehends	doesn't comprehend
considerate, thoughtful	inconsiderate, thoughtless
constructive	destructive, complaining
content-oriented	outer, surface, form-oriented
cooperative	uncooperative, unhelpful, combative
courageous	cowering, fearful
courteous	rude, impolite
creative	uncreative
deliberative	reckless
detail-oriented	scrimps on details

devoted	uncommitted, uncaring, hostile
directed, has direction	directionless, unfocused
disciplined	dissipating
does what is necessary, right	does what is convenient
perseveres, endures	relents, gives up
dynamic	passive
easy going	picky, fussy, choosy
enthusiastic	unenthusiastic, apathetic, indifferent
envious	envious, not
envisions the unseen	visionless
expansive	kept back, tight, constricting

faith in others	others can't be relied on
fatigue-free	tired, fatigued
flexible	inflexible, rigid, unbending, stubborn
focused	unfocused, scattered
forgiving	unforgiving, resentful, spiteful
freedom given to others	authoritarian, controlling
friendly	unfriendly, distant, aloof, hostile
frugal, thrifty	wasteful, spendthrift
generosity	stingy, miserly, selfish
goal-oriented	goalless, directionless
goodwill	ill-will, malice, hatred
graceful	clumsy
grateful	ungrateful, unappreciative

hard-working	lazy
high goals	low, no goals
higher social interests	lower, no social interests
honest	dishonest, deceiving, lying
humble	arrogant, conceited, ego-centric
idea-driven	ideas don't motivate to act
imaginative	unimaginative
improves self	stays the same
in rapidly expanding field of work	in static or declining field
initiates (has initiative)	lacks initiative
innovative	conservative
insightful	lacks insight, blind to, ignorant of
intelligence	stupidity
jealous, not	jealous, envious, covetous

kind	unkind, uncaring, cruel, mean
leadership	lack of leadership
modest	vain
objective	subjective, biased
observant	blind to, oblivious to
open-minded	narrow, close, small-minded
opportunist	
optimistic	pessimistic
organized	disorganized
patience	impatience
personable	non-engaging, distant, cold
persistent, sustaining	flagging, fleeting, unsustaining
polite, mannered	impolite, ill mannered, rude
positive	negative
practical	impractical, not viable
predictable	unpredictable

productive interactions with others	chit-chatting
punctuality	late, not on time
realistic	impractical
reliable	unreliable, undependable
respectful	disrespectful, rude, impolite
responsibility; takes-	blames others
responsible	unreliable, undependable
responsive	unresponsive, unreceptive
results-oriented	does for doing's sake, being merely occupied
risk-taking	averse to risk
seeing the whole picture	seeing only parts of the picture
seeking improvement	self-satisfied

self-confident	lack of self-confidence, insecure
self-directed	directed by externals
self-disciplined	undisciplined, unrestrained, indulgent
self-esteem, high	lack of self-esteem, lack of confidence
self-giving	self-centered
self-reliant	dependent
selflessness	selfishness
sensitive	insensitive, indifferent
serious	silly, trivial, petty
sincere	insincere, dishonest
social independence	social approval required
spiritual, inner connection	lacks any spiritual, inner connection
stamina	lack of stamina
stress-free, relaxed	stressed, tense
sympathetic	unsympathetic, unfeeling

systematic	unsystematic, disorderly, random
takes others point of view	insists on own view
thoughtful towards others	thoughtless, inconsiderate, callous
tough	weak, soft
trustful	suspicious, mistrusting
trusty	trusty not
unpretentious	pretentious, affected, ostentatious
unselfish	selfish
well-behaving	ill behaving
work-oriented	convenience first

Notes about personality traits: My friends tell me that I'm an organized person! For example, I always hang my keys on the key hanger. What's the difference between me and those who leave their keys wherever they land? It would certainly be more comfortable to leave the keys wherever I want. What happens in my mind when I leave the keys in a certain place? Actually, the fear of future suffering looking for lost keys is greater than the want of present comfort enjoyment. The reason that some people leave their clothes, books, and other belongings anywhere they land is that because the laziness joy (*want*) is greater than the visualized future *fear* of looking for their belongings.

What is laziness? The want of comfort is laziness. When you come home and change clothes, there is some suffering in hanging your clothes in the closet. So, you avoid the suffering and become lazy and leave all your clothes on the sofa or floor; however, you'll suffer looking for a particular shirt or shoe when you need it. As a matter of fact, for the organized person, the fear of future suffering is greater than the want of present laziness enjoyment. The visualized future suffering in an unorganized person is less than the want of laziness enjoyment. That's why an unorganized person doesn't return his or her belongings to their place.

Conclusion:

The degrees of personality characteristics (chemical settings of related genes) have the major role in our motions.

Chapter 13

HOW TO MAKE YOUR LIFE EASIER
Quick Meditation

Life could be easier if you accept its problems and discomforts. If you are outside and the weather gets cold, it's okay to suffer. You will enjoy the warmth when you get home. The more you suffer from the cold, the more you'll enjoy the warmth. You may say, how about if I catch cold. It's okay to catch cold and suffer. You'll enjoy when you get your health back. If you get a headache, it's fine. Later, you'll enjoy its absence. If there is no electricity tonight, that is good. You'll enjoy the power when it comes back on. If you cut your finger, that's no problem. You'll enjoy it once it's healed. If your car breaks down, that's all right. You'll enjoy the day it's fixed. When your teenager is late and hasn't come home on time, it's okay to worry. You'll enjoy it when she's back. When you get fired, it's okay to be upset. You will enjoy it when you find another job. When you're sick in bed for a few days, fine to suffer. You'll enjoy it when you get your health back.

Clearly, life could be easier if we accept its problems.

The problems are here to create activity and life for us; we will be like dead people without problems. We need to change our outlook. We need to believe that cold, envy, dishonesty, germs, illness, and so on, are necessary for life. We need to believe that problems bring us activity and happiness.

We see the world differently when we change our vision. All bad things will appear to be different in our minds, so when you see your child suffering from a fever, you won't lose control but will believe that it's okay to suffer. She will enjoy her health once she recovers. It's okay for you to suffer, too; you'll enjoy, too, once she gets her health back.

There is always joy waiting for us after each suffering

You can use this tool for quick meditation. Just remember there is joy for every problem you solve.

Anytime you are unhappy from any discomfort or problem, just think on the enjoyment waiting for you. As such, we should thank nature for our problems. Problems and discomforts bring suffering, which results in enjoyment. There is a joy for every suffering; enjoyment without suffering is impossible. That's how our nature is—we cannot have absolute enjoyment. If we have no problems or discomforts of any kind, we would remain motionless. People who have ultimate comfort and have nothing to work for or to challenge, often face severe depression.

Thus, meditation is nothing but releasing ourselves; by accepting the fact that there is always joy for any suffering, we can adjust ourselves from a high to a moderate level of suffering by believing without a doubt that there is joy waiting us.

Sometimes you may believe that there is no joy in your suffering. In such cases, you are paying off the joy you experienced earlier, so you can still modify your suffering to a lower level by accepting the fact that you already experienced the joy in the past, and now it's time to pay it off.

Example one:

You spend money from your credit card and have a good time on vacation. You're supposed to work (suffer) and save money in order to be able to go on a vacation (joy), but you went on vacation first (joy), and now you have to work hard (suffer) to pay off your debt. You cannot avoid suffering when you experienced joy first, but you can break them down to a longer period of time and suffer less at the time, exactly like a loan payment—you may take lower payments for a longer time instead of high suffering for larger payments in a short time.

Example two:

You suffer tremendously when one of your loved ones passes away. You can alleviate your suffering level by believing that you are paying off the enjoyment you had before this person's companionship.

If you use the above logic, you will find that you can handle problems much more easily, and you won't blame others for your own actions. The reality is that you must pay off your prior joys before you can expect a reward for suffering.

I'm sure that you will be able to lower your suffering sums by practicing this method of meditation, but remember when you lower your suffering sum, you lower the following enjoyment as well.

Conclusion:

We can lower our suffering by performing a quick meditation on the fact that enjoyment is waiting for us after each suffering, or that we are paying off the past enjoyment.

Chapter 14

HOW TO GET ALONG WITH OTHERS

Getting along with others is the greatest problem in today's societies and families. Most people in my community live with a dog, and family members are rarely happy living together. Why? Because we think people can control their emotions and they don't (chapter on Self-control), and because we expect people to use the power of their free will and make decisions regardless of their feelings, and they don't do it. There is nothing wrong with people's actions and reactions; they function perfectly and precisely according to their DNA affected by the surrounding environment, nothing more or less. The problem is that in our minds, we can't believe that there is no self-control. If we learn that humans have no self-control, then our vision and expectation elevate to a different level. Our tolerance would be much higher, and we would get along with people easier. The more we accept that there is no self-control, the more we could tolerate others' anger and nervousness. An analogy that I have in mind is to compare a car with a broken water pump with a short-tempered person. If you have a used car and you drive uphill, you expect the radiator to boil because of the broken water pump; thus, you don't get mad at the car when it is boiling. If your husband is short-tempered, and you have the knowledge that the related gene makes him boil easily without his control, then any time he boils, you handle him as if he had no control over his reaction. Subsequently, you develop your tolerance towards his nervousness.

We think that there is self-control and that we can control our feelings, but if we analyze the factors involved in each particular action separately, we will find that there is always another greater desire or another greater fear that overcomes the controlled felt subject. For example, a friend of mine put a hundred dollars bill in the fire to prove he has control over his feelings, and he explained: I love money, but I can act against my emotion and burn the thing I love. I responded: When I said humans don't have control over their feelings, I challenged your belief, and you lost your balance; then your brain looked for an action to regain your balance. Your desire to regain your lost balance was greater than the fear of losing a $100 bill in this scenario. Or we can say the suffering from imbalance was greater than the suffering from losing the $100 bill; thus, your brain chose the action with less suffering. That's why you burned the bill, not because of your power of controlling the feeling. Based on this example, there is always another greater want or fear that overcomes the controlled felt subject, and we think that we can control our emotions but we don't.

People may also try to control their emotions in others' presence to save the perception of themselves, but it is just a fake action. If you follow the person's emotions when nobody is around, you'll see that their actions are the reflects of their characteristics. Perception creates a greater fear that motivates individuals to act differently.

Focusing on people's outstanding characteristics and the repetition and uniformity of their actions and reactions leads us to the fact that there is no independent power such as self-control beyond human bodies.

The following characteristics and people's repetitive actions and reactions clarify how humans act with no self-control:

There is a repetitive and uniform pattern of actions and reactions in your envious friend.

There is a repetitive and uniform pattern of actions and reactions in your short-tempered boyfriend.

There is a repetitive and uniform pattern of actions and reactions in your stingy friend.

There is a repetitive and uniform pattern of actions and reactions in your disorganized co-worker.

And there is a repetitive and uniform pattern of actions and reactions in the people who are impatient, narcissist, picky, conservative, clumsy, pessimistic, and so on.

The above observation and the related experiments prove that **there is no power beyond our bodies such as self-control.**

Note from the author:

Dear reader,

I'm Eddie Rafii, the author of this book. I have hesitated to write recommendations, problem solutions, and life instructions in this book for the following reason:

Any advice or life instruction might alleviate your sufferings and thus lower your enjoyment sum as well by the law of suffer and joy in our nature. If I could find a way to elevate your enjoyment and keep your suffering low, I would definitely advise you to follow my life instructions. Unfortunately, there is no way to enjoy more and suffer less. The rule of nature discussed in the chapter on Happiness tells us that joys and sufferings are equal in normal conditions. Happiness, as a lifetime joy, doesn't exist. We must suffer to enjoy.

Please note that the subject discussed in this chapter and in the chapter on "How to make your life easier" are just suggestions, not recommendations.

I HOPE THIS BOOK DEVELOPS OUR MINDS TO THE EXTENT THAT WE GET ALONG WITH OTHERS MORE EASILY

Chapter 15

ARE MORALITIES HUMAN VALUES?

MORALITIES ARE NOTHING BUT FEARS OF SUFFERING.

Here is the analysis:

CONSCIENCE?

The definition of conscience from dictionary.com is "the complex of ethical and moral principles that controls or inhibits the actions or thoughts of an individual." Let's find out if this definition is true or not.

Example 1: You are the guardian of an orphan, and you'll get a bad feeling if you spend his or her money on yourself. Does conscience stop you from stealing the money or does the fear of suffering prevent you from the dishonest action?

Example 2: You are honest with your wife and now you have to lie to her because you've cheated on her. You'll have a bad feeling if you do so. Is this feeling moral or is it a fear of suffering?

Example 3: How would you feel if you want to seduce your friend's wife? Is there any other factor that prevents you from doing so, or is it the fear of suffering?

Defining conscience as human value is wrong definition. The fear of suffering prevents humans from performing wrong actions, so conscience is nothing but the fear of suffering. [F14]

DISCIPLINE?

I asked myself why I always place my mailbox key at a certain spot on the fireplace mantel. What's the cause of this discipline? After thinking about this act, I found that the cause of this action is fear, fear of suffering caused by having to look for the key when I need it again. Although it would be convenient to leave the key anywhere, so the fear of suffering drives me to place the key in a certain spot. Can we say that discipline is nothing but the fear of suffering? [F15]

PUNCTUALITY?

I really suffer when I'm late for my appointments. Does morality force me to be punctual or the fear of suffering? [F16]

SACRIFICE?

You sell your home to provide money for your child's medical needs. How would you feel if you didn't sell your home and your child died? You would suffer from your act, so the fear of suffering makes you sell your house, and the sacrifice is nothing but the fear of suffering.

RESPECT?

You respect your parents. You'll feel bad and imbalanced if you don't respect them; the fear of suffering from imbalance makes you respect your parents. Isn't respect nothing but the fear of suffering?

The above analysis opens a new dimension of thinking to see the realities of human nature:

Conclusion:

Are moralities human values? Or, moralities are nothing but the fears of suffering.

Chapter 16

MENTAL DISORDER OR CHEMICAL DISORDER?

Any emotional reaction is the result of certain chemical transactions. Mental disorder is improper terminology and should be replaced with chemical disorder or chemical imbalance.

MENTAL DISORDER

OR

CHEMICAL DISORDER

Dopamine
http://www.youtube.com/watch?v=o2T-7_g6yUU

Oxytocin
https://www.youtube.com/watch?v=_DXOpMCITw0

****

"There are medications today that may help in almost any emotionally connected issue: obsessive worry, panic anxiety, generalized anxiety, inattention, decreased memory, manic highs, depressive lows, premature ejaculation, impotence, decreased sex drive, too much sex drive, anger, insomnia, and many more".

"The mind and the body are one and should not be treated separately"

From research we know that chemicals in the brain (neurotransmitters) play a large part in the symptoms listed above. Psychiatric medications today can often help significantly in abating issues such as those listed above.

If dopamine is altered, then one may be out of touch with reality; if GABA is altered, one may be anxious; if Acetylcholine is altered, then one may have memory problems; if norepinephrine is altered, one may be manic high; and if serotonin is altered, one may be depressed, worried, more irritable, and have insomnia. Today we not only know which neurotransmitters produce certain emotions, we even know about neurotransmitter subsystems. For example, there are at least eighteen subtypes of serotonin receptors alone. We even know what the subneuro transmitters do, and we have medications that will not only affect specific neurotransmitters but also subsystems. The biochemistry of emotions has become a science in and of itself."

The Minirth Clinic P.A.

CHEMICAL IMBALANCE
OR
MENTAL ILLNESS

Mental illness or Chemical imbalance
http://www.youtube.com/watch?v=LLUoG9Se77w

There is nothing beyond the body such as mental illness. Abnormal actions and reactions are just the results of chemical sets that differ from the norm. Normal and abnormal emotional behaviors have a similar root in physical behavior. Normal and abnormal emotional actions and reactions are the results of a series of chemical processes; action will be abnormal if there is an abnormality in the produced chemicals in the brain.

Gene properties play a major role in one's behavior. Gene properties are based on related chemical structures. Abnormal behaviors are a reflection of the abnormal structure of genes.

https://www.youtube.com/watch?v=JQEiux-AOzs

Mental disorder is an improper term for abnormal behavior and should be replaced with chemical disorder.

Conclusion:
Every emotional feeling is the result of certain chemical processes in the brain. There is nothing external. [F13]

Chapter 17

BRAINWASHING AND BALANCE

Brainwashing is nothing more than generating an imbalance in others. Terrorism is the first thing that comes to our minds when we talk about brainwashing. Experts who train terrorists do nothing but create an imbalance in their trainees; this imbalance creates a want in the trainee. A well-trained terrorist has a great *want* for his task and no *fear* can stop him—even death. The only way to regain his balance is to complete his mission.

Military training is another common form of brainwashing. People who join the military have no desire to kill other humans. It is the training that effectively leads soldiers to drop bombs and kill thousands of human beings—not only without guilt but with a sense of accomplishment. Why? Because the only thing that can restore their imbalance is a successful mission.

We all are brainwashed. Television is the main source of brainwashing—through advertising and the promotion of celebrity status (the idea that if we use the same products as celebrities, we will achieve their status). Today is October first-2008. These days, large sunglasses are fashionable. Your friend will laugh at you if you wear sunglasses you bought eight years ago when small sunglasses were fashionable. Eight years ago, your friends would have laughed at you had you worn your large sunglasses from ten years ago. Moreover, it's prestigious to wear brand names with the logo displayed across your chest. Unknowingly, we advertise products and, instead of getting paid for such advertising, we pay them high prices for their products—that's good brainwashing.

My ex-wife believed that the most expensive skin creams are the best. She used to get mad when I asked her to use one on only one side of her face for thirty days to determine if it's truly effective. I conducted just such an experiment. I bought a product that promised to reduce wrinkles and applied it to only the right side of my face. After thirty days, I asked my ex-wife if she could tell which side I used the product on. She looked carefully and thought for a while—a sign that she couldn't tell the difference. She then pointed the left side of my face.

I always ask myself why she didn't want to conduct this same experiment, and why she got mad when I asked her to do so. The only answer I have is that the cosmetics companies have done a great job of brainwashing her.

The advertisers create such an imbalance in us that we will pay an entire month's salary for a brand-name purse or piece of clothing.

Television is the center of this brainwashing, as advertisers create an imbalance in us. We buy their products and soon find most of them stored away in our garages.

Conclusion:

Brainwashers generates imbalance in us. The brainwashed person makes motions in favor of the brainwasher in order to regain his or her balance. [F25]

Chapter 18

DO WE REALLY DO THINGS FOR OTHERS?

You see a poor man in the street and give him some money. You miss your daughter so you visit her. You give a gift to your friends for their new home. If I asked you why you acted as you did in these situations, you would say because you wanted to make the man/your daughter/your friends happy. What would you say if I told you that you did so to make yourself happy?

Example one:

You lose your balance when you see a man suffering from hunger, so you help him to regain your balance. How would you feel if you ignored him instead? I'm sure you would feel bad and imbalanced. The level of your comfort wouldn't be the same as it was before the event. In other words, you suffered from losing your balance. Your body made a motion (giving money) to regain your balance, and, as a result, you enjoyed regaining your balance (fact four); you helped the poor man to get rid of your bad feeling.

Example two:

You miss your daughter so you go to see her. What would you feel when you miss her, but you don't go to see her? It's obvious that you would suffer. Therefore, you visit her for your own enjoyment.

Example three:

Your friends bought a new house, and you give them a gift. What would your feeling be if you visited them empty-handed? Of course, you would suffer. You give them a gift for your own enjoyment.

Thus, our motions serve only ourselves.

Let's analyze example three in detail. Losing your balance causes you to buy a gift. This could result from one of the following (or a combination of) factors:

(1) You love your friends, and when you give them gifts, they become happy. Your enjoyment stems from their happiness. You give them gifts for your own enjoyment.

(2) It's customary to give gifts when someone buys a new house. If you visit without a gift, you will lose your balance and suffer when people think that you're cheap. You give a gift to prevent yourself from suffering (fear of suffering from being perceived as cheap).

(3) Your friends brought you a gift when you bought your new home. You'll remain imbalanced if you visit them empty-handed. You give a gift to prevent yourself from suffering. Any other factors not listed here generally fall into one these three categories or can be

analyzed in a similar manner.

Let's analyze these three factors one by one. Factor one (love): You enjoy making others happy. Factor two (perceived as cheap): The fear of suffering from the thought of being cheap makes you give a gift. Factor three (perceived as careless): The fear of suffering from the thought of being careless makes you give a gift.

Regardless of which factor causes you to lose your balance, you'll suffer if you don't give your friends a gift. Thus, you give a gift for your own enjoyment or to prevent suffering.

Example four:

You sacrifice and sell your home to provide money for your child's medical needs. How would you feel if you didn't sell your home and your child died? You would suffer from your act, so the fear of suffering makes you sell your house. So, sacrifice is nothing but a fear of suffering.

Conclusion:

We do things only for ourselves. There is no sacrificing in reality. [F21]

Chapter 19

INSTINCT AND FREE WILL

What is the difference between instinct and free will? When you see a car hit a child and the child dies, you cry. In doing so, are you following your instinct or choosing to cry? You lose your balance, and you cry to regain it.

You see a mouse, and you scream. Do you scream by instinct or do you choose to scream? You see mouse, you get scared, and lose your balance, so you scream to regain your balance.

Someone gives you a flower, and you smell it. Do you smell it according to your instinct or do you choose to smell it? You see the flower and become curious about its smell. You lose your balance from such curiosity, so you smell it to regain your balance.

Someone gives you a flower, and you say, "Thank you." Do you say thank you by instinct or choice? When you receive the flower, you feel uncomfortable if you don't say thank-you; doing nothing will result in remaining imbalanced, so you say thank-you to regain your balance.

You turn on the light when you enter a room at

149

night. Do you do so by instinct or do you choose to do so? You lose your balance because of the darkness, so you turn on the light to regain it.

You promised your daughter to take her to a movie this afternoon, but your mother calls and says that she doesn't have a ride to her doctor's appointment. What would you do? Does your instinct tell you what to do, or does free will make the decision for you? (Are you free to choose your action?) In this scenario, you face the fear of suffering from your daughter's state of upset if you don't take her to the movie, as well as your fear of suffering from your mother's anger if you don't give her a ride. You will do the motion that involves less suffering.

No matter what our belief in free will or instinct is, the mechanism of each action or reaction is based on our balance and the potential outcomes of our wants and fears (facts one, two, and seven).

Fact one: We always want to be in balance.

Fact two: When a factor affects us, we lose our balance and make motions to regain our balance.

Fact seven: The degree of our characteristics and the environmental conditions, determine the levels of our wants and fears, and the sum of the outcomes of our wants and fears determine our motions.

Have you ever been in a situation in which you wanted to jump over a stream but were not sure if you could make it? If you try to remember the involved factors, you'll find that there are usually two major factors; your *want* to jump and your *fear* of failure. I'm confident, that the sum of the outcome of your want and fear determine your motion.

We frequently see cats that want to jump from one point to another. Next time, when you see that a cat wants to jump, watch carefully and see what happens. One factor—for example, food on the other side— creates the want, while the distance of the jump creates the fear of failure. The cat will stop and look carefully. If it isn't that hungry (a low level of want), it won't jump; but if it is very hungry (a high level of want), it will jump. The fear factor is fixed, but the want factor could vary depending on the condition; the sum of the outcome of want and fear determines the act's action.

Humans and animals' motions are identical

As we see, the function of humans' and animals' motions is identical. Let's look at another example.

If you ever tried to feed pigeons from your hand, you have seen that they come around you, but are afraid to get too close to you. Here, there is a *want* to eat and a *fear* of being caught by you. The pigeons perform motions according to the outcome of want and fear.

Want to eat and *fear* of being caught

A hungry person who needs food goes to a market but doesn't have enough money to pay for the food. Here, there is a *want* to eat and a *fear* of being caught for stealing food. The person makes motion according to the sum of the outcomes of his want and fear. He won't steal if his fear is greater than his want, but he will steal if his want is greater than the fear.

Conclusion:

The mechanism and functionality of humans and animals' motions are identical, [F20], so why do we believe that animals act by instinct and humans act according to the free will?

Chapter 20

HOW WE TAKE THINGS FOR GRANTED

Why do we get used to things? We buy a painting that we love very much and hang it on the wall in our living room. We continue appreciating it day by day. Weeks or months later, we find ourselves forgetting that the painting is even hanging on the living room wall.

Buying a new car is another example. We take pleasure for the first few days—if not months—but over time we get used to having the car, and it's no longer as exciting driving it as when we first bought it.

Marrying the ideal spouse is the same. We appreciate his or her company and feel so lucky to share a life with this person, but eventually, the excitement dies down, and we become used to his or her presence.

Example one:

You see a beautiful painting in an art store, and suddenly you lose your balance. You want to have it, but you can't afford it. The next week you happen to pass by the store and see the painting again, so you purchase it. You are careful not to damage it as you carry it in and out of your car. Once you have it in your home, you'll only enjoy it until you have seen it so many times that it no longer seems as interesting. Why? You lose your interest when the total enjoyment of the painting equals the suffering you faced before acquiring it. You may still enjoy having the painting but not necessarily looking at it; you may appreciate it less once you own it. If a friend comes over and is moved by how beautiful the painting is, you will feel good that you own such a piece of work or enjoy the feeling of being wealthy enough to own an expensive work of art. These forms of enjoyment differ from the enjoyment of the art itself. The desire to be wealthy and successful allows you to feel superior to others. You gain a sense of power at times when people feel envy toward you. Such enjoyment depends on the characteristics of our personalities and differs from the enjoyment of art. The enjoyment of owning (or showing off) the painting is totally different than the enjoyment of looking at it artistically. We take a beautiful painting for granted when the total joy of looking at it pays off the total suffering of acquiring it.

We take things for granted every day. Some that I have in mind are our vision, spouse, children, health, freedom, peace, and so on. When we lose our health, we suffer during the illness and particularly enjoy the day we regain our health. The daily enjoyment of health gradually decreases until it reaches zero, which is when the total joy pays off the total suffering. At this point, we start to take our health for granted again.

We take our peaceful lives for granted. If we experience war and chaos, we will suffer and subsequently enjoy life to a high degree once peace is restored. Our enjoyment from peace lessens daily until we no longer enjoy peace because the total enjoyment of peace equals the total suffering from war. Consequently, we take peace for granted again.

At this point, let's reexamine the research about lottery winners.

****

An interesting study (Brickman, Coates, and Janoff-Bulman. The researchers studied both lottery winners and individuals that sustained a physical injury, to determine if winning the lottery made them happier or if sustaining an injury made them less happy. What they found was that immediately after either event, levels of happiness were higher (lottery winners), or lower (physically injured), and that after eight weeks or less, people returned to the level of happiness they had *before* the event. This research suggests that we adapt to these situations very quickly, and often return to the degree of happiness we had *before* such an event.

Researchers (Brickman and Campbell (3)) argue that all individuals labor on a "hedonic treadmill." As we rise in accomplishments and possessions, our expectations also rise. Soon we get used to the new level and it no longer makes us happy. Has this ever happened to you? Maybe when you bought a new car? Only to find out that what you really wanted was the feeling the car would bring you, not the car itself. These are just a few of the difficulties in understanding what we mean by happiness.

Source: University of California Regents
Human Resource
HR Update-February 2002

****

Initially, a lottery winner has a lot of joy because of his or her wealth and what he or she can do with it; however, the joy derived from his winnings gradually decreases until it reaches zero—when the total joy pays off the suffering before winning. At this point, the winner starts to take the new quality of his or her life for granted.

We need to miss our loved ones to enjoy their presence. We need to lose our health to appreciate it. We need to suffer from war to enjoy peace. Therefore, we need to suffer in order to enjoy.

Conclusion:
We get used to things when the enjoyment counteracted the suffering and vice versa. [F23]

Chapter 21

HOW WE ADAPT OURSELVES

We adapt ourselves to the new conditions of life. When something pleasant or unpleasant happens in life, we adapt ourselves to new situations. For instance, when one of our loved ones moves to another city, we miss him or her in the beginning, but eventually, this feeling subsides. When one of our loved ones dies, we initially miss him or her terribly, but eventually, such feelings become less and less intense—we may even forget them. When you earn your diploma, you are very happy, but eventually, you don't feel the same level of happiness. When you buy your dream house, you are very excited, but after a while, you don't feel the same. Why do our feelings change over time?

When the total suffering pays off the total enjoyment, we adapt to the new situation. In addition, when the total enjoyment of happiness pays off the total suffering, we adapt to the new condition. To clarify, let's analyze the above examples.

When one of our loved ones moves to another city, we initially suffer quite a bit. Our suffering gradually decreases until we don't suffer anymore; at this point, the total suffering has paid off the total enjoyment of the relationship. Consequently, we have adapted to the new situation. The same is true when a loved one dies. At first, we suffer a lot; however, our suffering lessens day by day until we don't suffer anymore, at this time the total suffering pays off the total enjoyment of past relationship, and we adapt to the new condition.

When you earn your diploma, you are initially excited by your accomplishment. Your enjoyment decreases day by day until you don't enjoy it anymore; at this time the total enjoyment pays off the total suffering stemming from the studying and hard work, so you have adapted to the new situation. When you buy a new house, you enjoy it a lot in the beginning. Your enjoyment gradually lessens until you don't enjoy it anymore. The total enjoyment pays off the total suffering from the hard work and you adapt to the new situation.

The following research supports these analyses.

****

An interesting study by Brickman, Coates, and Janoff-Bulman:

The researchers studied both lottery winners and individuals that sustained a physical injury, to determine if winning the lottery made them happier or if sustaining an injury made them less happy. What they found was that immediately after either event, levels of happiness were higher (lottery winners), or lower (physically injured), and that after eight weeks or less, people returned to the level of happiness they had before the event. This research suggests that we adapt to these situations very quickly, and often return to the degree of happiness we had before such an event.

Source: University of California Regents
Human Resource

HR Update-February 2002

Happiness: Something to Think About

7/14/2005

Conclusion:

We adapt to new situations when the suffering counteracted earlier enjoyment. We also adapt when the enjoyment counteracted earlier suffering. [F23]

Chapter 22

HAPPINESS ANALYSIS

Most people think that wealth is the key to happiness. Others believe that they feel the happiest when life is free of problems. Some say happiness is peace of mind throughout their lives. However, many wealthy people live in this world. Are they all happy? Many people live with almost no problems. Are they all happy? Could a person have peace of mind at all times? Have you ever seen a person who is constantly joyful throughout life?

What is happiness? I have asked this question of many people and received just as many answers. Most don't know how to describe happiness. For me, the definition of happiness is the feeling of joy. You are happy when you have this feeling and become unhappy when you lose it, so can a person be happy all the time? Can a person enjoy life at all times?

As I discussed in chapter I, joy cannot exist without suffering. "Joy and suffering always accompany every motion". Fact four.

Note:

Some sufferings and enjoyments in our daily life are not involved with the motions. For example, we may suffer the loss of our favorite team in a basketball game or from seeing dead people in a news story about an earthquake. We may enjoy watching a comedy movie or observing beautiful flowers; we might suffer if we see a dead bird while walking in a park or experience joy when seeing that a friend bought a new car. We do not initiate any action when this kind of joy or suffering occurs. I believe that these types of suffering and joy may not be counteracted in the same day but will be counteracted over a longer period of time in a normal life.

Problems bother us and create an imbalance in us. We make motions to regain our balance. There is suffering and joy always accompany every motion and life is nothing but repetition of losing and regaining balance. In the other words, life is nothing but the recurrence of suffering and joy. **If we eliminate problems from our lives, we eliminate joy and happiness as well. [F24]**

If no problem and discomfort of any kind exist, we get bored and suffer from boredom. When the boredom continues, we start to suffer from depression. It seems as though it is a never-ending cycle that we have to suffer one way or another. Work is the safest kind of suffering, work is stressful and makes us tired, nervous, and anxious. Work is suffering and that suffering not only gives us the joy of rewards, but it's also necessary for our mental and physical health. People who don't work and have ultimate comfort and seem to have abolished suffering in their lives will sooner or later lose their physical and mental health if there is no other serious activity involved. Exercise and similar practices seem to be a fun way to replace working, serious exercises also involve suffering.

Those who don't like their work enjoy more than those who like their work when they get off work, because they suffer more at work.

We always want to enjoy life, but joy without suffering is impossible—that's human nature. When you go to bed and before falling asleep, think about tomorrow and instead of expecting a nice day, expect a hard day with suffering. When you wake up, be prepared for a bad day. Don't be worried about having a bad day, though, because joy and pleasure are waiting for you anyhow. We need to reverse our thought and expectation to see how happy we are. **Can we say life is fair when our enjoyment equals our suffering?**

Can we say life is fair
when our enjoyments equal our sufferings?

The degree of our expectation plays a big role in determining the level of our happiness. The least happy people are those who have the highest expectations. Accepting problems and expecting suffering prepare our minds for low expectations, which is how we can alleviate our suffering and feel relaxed.

We can learn how to accept problems by practicing. Before practicing, we need to learn about our nature and believe that our nature needs problems, just as our bodies need food and water. What do we do if there is no food? We look until we find some. The same is true of problems. Our nature needs problems. If we have no problems, we have to look for some. Our minds are determined to look for joy and happiness. But to find them, we must first look for problems to experience suffering.

You might be surprised to learn that most arguments between couples and friends are based on this reality. The next time you are in the presence of friends or spouses arguing, focus on the cause. In most cases, you will see that one person is looking for a problem. It's true that our nature needs problems; we look for them when we don't have any.

Similar cases are evident among children. When they are unhappy, they pick quarrels and bother you. They have no answer when you ask them what's wrong. You try to find the cause of their unhappiness, but you can't find it. Why? Because you never considered the fact that their nature also needs to suffer—you always want to see them happy. You give them everything, but there is nothing left that you can do. The truth is they are bored and looking for problems. Parents unknowingly taunt their children, asking them "Are you looking for trouble?" The reality is yes, they are.

I should say that more or less every single person looks for problems. A person who climbs Mt. Everest goes through lots of problems and faces dangers for the joy of conquering the summit. The person who starts to build a custom house for his family voluntarily suffers a series of problems. When you plan a picnic, you look for a bunch of preparation problems. You buy yourself problems when paying for a ticket to watch a scary movie. When you want to have a nose job, you buy yourself lots of pain and surgical problems. And you look for long-term problems when you want to have a child.

We pay to get scared. Don't we love to be bothered?

Why do we look for problems? Because of the joys that are followed by the sufferings.

Now that you know the cause of most arguments, you know how to handle your spouse or friend the next time an argument occurs. The person who has lost his or her balance because of boredom or another factor wants to restore his or her balance by arguing or fighting with a close, available person such as a spouse, sister, or friend. In most cases, the person apologizes as soon as he or she regains balance.

I remember the night that, my ex-wife and I had a nice walk along the beach; when the fun was over, she asked me if I wanted to go to a movie. I agreed, but the movie had already started by the time we got there. We decided to walk along the shopping center near the theater instead. We started talking about our lives, and my wife mentioned that sometimes she didn't feel happy. I asked her why, but she remained silent. Her silence told me that she didn't know the cause of her unhappiness.

Before I continue, I would like to share a few words with you about our relationship. She is a very nice and beautiful lady; she has a good heart and moderate desires. I don't want to describe my own personality because I might not see myself as others see me, but I will tell you my ex-wife's words about me. We were married for six years; during this time, she has told me that I am almost a perfect man on four occasions.

As we walked in the shopping center, I started talking after a few seconds of silence. I told her that there are a few reasons that she didn't feel happy. First, I said that I'm not a normal man. A normal person argues, which turns into a fight; consequently, couples don't talk to each other and suffer, followed by their enjoyment when they make up. This cycle repeats throughout the relationship. However, my ex-wife and I used to miss the joy after the fight because there were no fights to suffer through.

Second, I said she feels unhappy because she doesn't have any responsibilities. If she likes to cook, she cooks; if she doesn't, she doesn't cook. The same is true of cleaning. If she wants a job outside the home, she can work; if she doesn't, then she doesn't work. She has no pressure. She feels no responsibility for the house mortgage or bills. She does not suffer from a lack of food on the table. She does not feel insecure because her husband is not a gambler or drunkard. In a nutshell, she doesn't feel happy because she has no problem and no suffering in her life.

She remained quiet, then looked at me and said, "Maybe if you would fool around, you could bring some excitement to my life." Her words were so surprising, but it confirmed that she not only understood the cause of her unhappiness but also looked for trouble through which to suffer to ensure her upcoming happiness.

Accepting reality is the first step toward alleviating suffering. Analyzing your nature and recognizing its needs and how it works will lead you to open your

mind to see the problems with unique insight.
An unfiltered vision of life and happiness will create a new dimension in which we can truly accept problems. Then we will experience a life in which most problems bother us less than before.

Happiness is temporary joy followed by suffering and terminates when the suffering is counteracted. Money, wealth, and technology may bring us comfort and temporary happiness, but this happiness ends when the earlier suffering is counteracted, and we feel unhappy again when we get used to the comfortable life.

Here I would like to bring to your attention that on some occasions, happiness comes first, and suffering shows up later, such as winning the lottery. You may object to this analogy: the person who wins so much money spends it and enjoys it. Where is the suffering?

The happiness from such money is much less than happiness from money earned through hard work. Clearly, hard work results in long periods of suffering, but such suffering is like saving happiness in a savings box—you drop a piece of happiness in this box every time you suffer from work. Happiness builds up in the savings box; the more you suffer, the more happiness you save for the future. You can enjoy as much happiness as you saved—nothing more or less. As a matter of fact, your happiness ends when the suffering is paid off.

Lottery winners experience a sudden, significant change in their lifestyles without having worked for it. Their expectations increase and they spend money easily. Sooner or later the level of their happiness drops; they get used to the quality of their life and don't enjoy it as much as they first did. The level of their happiness returns to the level before winning, by getting used to comfortable life.

****

Lottery winners, trust-fund babies and others who get their money without working for it do not get as much satisfaction from their cash as those who earn it, a study of the pleasure center in people's brains suggests.

Emory University researchers measured brain activity in the striatum—the part of the brain associated with reward processing and pleasure—in two groups of volunteers. One group had to work to receive money while playing a simple computer game; the other group was rewarded without having to earn it.

The brains of those who had to work for their money were more stimulated.

"When you have to do things for your reward, it's clearly more important to the brain," said Greg Berns, associate professor of psychiatry and behavioral science.

"The subjects were more aroused when they had to do something to get the money relative to when they passively received the money."

Real-World Implications:

Berns and other researchers said the study has broader real-world implications, particularly in the age of multimillion-dollar lottery jackpots.

He said that other studies have shown "there's substantial evidence that people who win the lottery are not happier a year after they win the lottery. It's also fairly clear from the psychological literature that people get a great deal of satisfaction out of the work they do."
Source: Associated Press
Are Lottery Winners Really Less Happy?
msnbc 7/10/2005

****

According to Dr. H. Roy Kaplan, author of several books on lottery winners, "winning the lottery doesn't change people's lives as much as is imagined."

****

'Money doesn't change a person's level of happiness," said Kennon Sheldon, a psychologist at the University of Missouri at Columbia. "We consistently find that people who say money is most important to them are [the unhappiest]," Sheldon said.

****

Nearly one-third of lottery winners become bankrupt.

The CFP Board made an offer to the National Association of State and Provincial Lotteries to provide the organization's members with information to distribute to winners. The *Investment News* article highlighted the lack of financial guidance many winners receive from state lottery agencies; estimates show that nearly one-third of lottery winners become bankrupt.

****

A new study by American psychologists has found that cash and popularity does not bring nirvana. Experts say that excessive wealth, particularly for people unaccustomed to it, such as lottery winners, can actually cause unhappiness.

There is evidence that there are very wealthy people who are very unhappy, particularly people who were not born to wealth like lottery winners.

Source: BBC News
2-12-2001, 00:05 GMT

Money Can't buy Happiness
http://news.bbc.co.uk/2/hi/health/1162153.stm
7-17-2005

Happiness of lottery winners:

CNN:
http://www.youtube.com/watch?v=Aom1eSyI6F4

174

FOX NEWS

England:
http://www.youtube.com/watch?v=gyn2-HRLDNo

****

Happiness plays an important role in our well-being. So, what makes us happy? Objective life circumstances (am I living the good life?) account for only 8-15 percent of happiness ratings according to one researcher (Lyubomirsky (1)). Maybe we would be happier if we won the lottery? An interesting study (Brickman, Coates, and Janoff-Bulman (2)) suggests otherwise. The researchers studied both lottery winners and individuals that sustained a physical injury, to determine if winning the lottery made them happier or if sustaining an injury made them less happy. What they found was that immediately after either event, levels of happiness were higher (lottery winners), or lower (physically injured), and that after eight weeks or less, people returned to the level of happiness they had *before* the event. This research suggests that we adapt to these situations very quickly, and often return to the degree of happiness we had *before* such an event.

Researchers (Brickman and Campbell (3)) argue that all individuals labor on a "hedonic treadmill." As we rise in accomplishments and possessions, our expectations also rise. Soon we get used to the new level and it no longer makes us happy. Has this ever happened to you? Maybe when you bought a new car? Only to find out that what you really wanted was the feeling the car would bring you, not

the car itself. These are just a few of the difficulties in understanding what we mean by happiness.
　　Source: University of California Regents
　　Human Resource
　　HR Update-February 2002
　　Happiness: Something to Think About
　　7/14/2005

****

Here I would like to analyze how the level of happiness of lottery winners returns to its level before the event in a short time, as well as how the level of the unhappiness of sustained injured returns to its level before the incident.

A-The analysis of the lottery winner's happiness:

We all have the desire for a bigger house, nicer car, quality clothes and so on, and we suffer when we compare ourselves with rich people. It is obvious that the level of happiness of lottery winners jumps to a high level as soon as they win and experience the joy of buying and doing things that they always wanted. The level of their happiness returns to the level before winning as soon as the total enjoyment counteracts the earlier sufferings.

B-The analysis of the unhappiness of those who sustain a physical injury:

If we observe the children before they start walking,

they have a great desire to be able to walk, and they suffer by watching others walking and realizing that they can't. Sooner or later, they develop the ability to stand on their feet, and doing this is so enjoyable for them that they try harder and harder until they can walk normally. Their total enjoyment of the whole process could be divided in two portions. The first portion pays off the total suffering before walking, and the second portion creates a higher balance in them that will be counteracted by the sufferings when they get old and lose their ability to walk.

Now if a man loses his legs in an accident before he gets old, losing his walking ability, he would suffer a lot in the beginning, but these sufferings get less and less day by day to the point that he gets used to the new situation. At this point, his total suffering paid off the second portion of the enjoyment when he learned how to walk. On the other hand, the level of his happiness returns to the level of his happiness prior to walking when he was a child.

The bottom line is; that there is no way to enjoy anything without suffering. Happiness, as a feeling of joy throughout a lifetime, it's an unrealistic believe. Happiness is a temporary feeling of joy that follows suffering and ends when the suffering is counteracted. We must suffer again to feel happy.

Since happiness requires suffering, does it make any difference if we have a high-quality life or not? Does it make any difference whether we are rich or poor?

I was invited to a fancy wedding last week at a five-star hotel. I was very hungry during the ceremony. Afterward, I was looking for something to eat and walked to a patio overlooking the ocean and found some bread and cheese. I helped myself three times, and I enjoyed it very much. The dinner was served almost one hour later. The salad was very good, the steak and fish were prepared well and the service was first class. However, I ate very little and didn't enjoy the food much. The enjoyment of a simple snack was so much greater than the lavish dinner. Why? I had suffered so much from hunger before the simple snack that I didn't suffer as much before the lavish dinner. As a result, the quality of the food played no role in my enjoyment; indeed, it was the sum of suffering that made the food enjoyable.

It doesn't matter whether we are poor and have a low-quality life or rich and have a high-quality life. It matters how much we suffer. The more we suffer, the more we feel joy. It's an unrealistic belief in the back of our minds that prevents us from feeling the real joy of life. We always have this incorrect belief that we have to be rich with a life free of problems to be happy. Yet without problems, we would be like dead people. Our nature needs problems to be motivated, and solving problems brings joy and pleasure. Life is nothing but repeating cycles of suffering and enjoyment—every day, every hour, and every minute.

To have a better understanding, let's compare ourselves with past kings. We will examine the quality of life of an average person in our time to that of a king from 500 years ago:

(1) Transportation: A king was transported in a horse or carriage. During such travels, he suffered from heat in the summer and, if he became thirsty, there was no water or only warm water that his servants carried for him. Today, the average person uses a personal car, carries a cooler with drinks in the trunk or stops at one of the millions of convenience stores, plays his favorite music, turns on the air conditioner, and travels in comfort to his destination.

(2) Communication: If a king wanted to communicate with someone in another city, he had to send a messenger to that city and wait for his return. This could take weeks. Today, the average person can speak directly with anyone practically anywhere in the world using mobile phone or instant messaging.

(3) Health care: The difference here is so obvious

that I don't even need to explain. A king could die from any simple disease, while today's average person has many medical facilities available.

(4) Food: Today's average person can choose what he needs from among thousands of items in a grocery store, including fruits and vegetables that are out of the season. Meanwhile, a king's choices were severely limited by what his vassals had killed hunting, what was harvested from the garden, and what could be preserved.

(5) Entertainment: Kings had to rely on their courtiers and traveling shows, which were not scheduled. In addition, if the skills of these entertainers were poor, the king may not have had the option of finding a new act. Today's average person can access a variety of television programs, concerts, comedy shows, movies, nightclubs, circuses, zoos, water shows, magic shows, and so on. If the quality is not up to their expectations, other options are readily available.

There are hundreds of other items that we can use as examples to compare the quality of our lives with past kings, such as clothes, shoes, safety, and sports. The quality of our lives is so much better that we can't even compare them, so we must be a hundred times happier than past kings—but are we?

If we compare the quality of our lives with that of an average man 500 years ago, we will see an even greater difference, so much so that comparison would be futile. Are we happier than humans who lived at other times?

The answer is, we are neither happier nor unhappier, obviously, humans in the past suffered and enjoyed more. Therefore, the quality of life doesn't bring us happiness; it's the suffering that brings us joy and happiness.

Being happy for a lifetime—a goal for many people— simply does not exist and is the wrong definition of happiness. Happiness is a temporary feeling of joy. For every joy, there is suffering (Fact four). Without a doubt, poor people suffer and enjoy more than rich people. Yet absolute enjoyment doesn't exist.

Every person generates millions of motions between birth and death. Suffering and joy exist in each motion. It seems the total suffering equals the total joy, measured when we are at zero balance. An analogy is mountain climbing. When mountain climbing, we walk up hills and downhills. Your total ascents equal your total descents when you get back to your starting point.

Our sufferings equal our enjoyments
when we regain our balance

Our suffering and joy are equal only when our balance returns to its prior level. For example, if we don't eat for a few days and then we eat a little, obviously, our suffering is more than our joy, but when we eat enough and regain our prior balance, then our suffering and joy are equal.

Note: certain events in our lives cause us deep suffering and no motions can make us regain our balance. In such cases, our personality changes. For instance, when we lend money to a friend and he or she doesn't return the borrowed funds, our trust changes and our balance never returns to its prior level. As a matter of fact, we aren't the same person; we are a new person with a changed degree of trust. Consequently, in such cases, our suffering is greater because we don't regain our balance.

The quality of life has no role in our happiness. In other words, it doesn't matter how we live, but how much we suffer. The more we suffer, the more we enjoy. There is no gain or loss, **so what are we doing? Why are we so hungry for wealth? Why do we lie, cheat, and steal?** For every lie, there is enjoyment and suffering. For every theft, there is enjoyment and suffering, and for every act of dishonesty, there is joy and suffering, so what do we gain? Are wealthy people who gain their money by stealing happier than others? They may suffer less, but they enjoy less as well.

If your rent is due if your car needs new brakes if you are not happy with your tiny apartment if you're worried because your child has a fever, or if the envy of your coworker bothers you, you are so lucky to have problems. Problems motivate you; anytime you solve a problem, you receive enjoyment. You don't want a life free of problems. **Ultimate comfort doesn't bring us the stable happiness.** Life without problems makes you inactive and depressed. Our nature needs problems to be motivated.

If you're living in a nice house and drive your desired cars if you're healthy and everyone in your family is the same, if your housekeeper does all the work, if you are not jealous of your friends' jewelry or furniture, if you're not worried about income, if the leaking faucet does not bother you and there is always someone there to take care of it, if you have perfect comfort and your life has been like this for a while, you're an unhappy person—or soon will be. This ultimate comfort we all pursue brings us unhappiness and depression. **Ultimate comfort is poison for life** that's why there is more suicide among wealthy and popular people.

Here are some studies about suicide.

****

Are suicide rates higher among 'the rich and famous' than among average people?

One case control series reported in BMJ in a relatively large sample (~1500) found that "people with a history of mental illness and a high income are at greater risk of committing suicide than their lower income counterparts. This may be because richer people with a mental disorder feel more shameful, vulnerable, and stigmatized."

http://www.bmj.com/content/322/7282/0.7.full
http://news.ycombinator.com/item?id=2157406

****

Friday, 9 February 2001, 00:54 GMT
Suicide risk for rich mentally ill

Rich people with a history of mental illness are up to three times more likely to kill themselves than those with less money, research suggests.

Scientists in Denmark have discovered that contrary to popular opinion it is not the poorer of society who are most at risk.

http://news.bbc.co.uk/2/hi/health/1160222.stm

****

According to Suniya Luthar and Bronwyn Becker at Columbia University, "research over the last 30 years indicates that Americans are twice as rich now but no happier than they used to be. Divorce rates have doubled, the suicide rate among teens has tripled and depression rates have soared."

http://www.crisiscounseling.com/Articles/PrivilegedKid
sAtGreaterRisk.htm

****

Why don't I feel happy?
Do I know my real nature?

Problems create wants, and wants generates motivations. Joy is the reward for solving problems.

Happiness and unhappiness are like any other phenomenon in nature: if we don't experience unhappiness, we wouldn't have any clue what happiness is. We easily understand that if we never experienced cold, we wouldn't know what the warmth is, but we all expect to achieve lifetime happiness without suffering— that's impossible. We need to remove the filters from our vision and see our nature as it is. We need to accept unhappiness so that we may feel the joy of comfort.

Conclusion:

Happiness as a lifetime feeling of joy is an unrealistic belief. Absolute joy doesn't exist. Happiness is temporary joy followed by suffering, and terminates when the suffering is counteracted. We must suffer to enjoy. The more we suffer, the more we enjoy.

Chapter 23

DOES TECHNOLOGY BRING US LIFETIM HAPPINESS?

When I was a child, our house wasn't wired for electricity. I remember doing my homework by the light of an oil lamp. We didn't have a car, so I walked to school, often getting wet in sudden rains. No effective pain relievers existed, so I suffered through any headaches I had. My toys were limited to a rubber ball and a few other simple toys.

Technology has made it possible for today's children to use numerous of items powered by electricity besides lighting, as well as have access to conveniences such as riding to school in a car or bus. Weather forecasting technology predicts sudden rainy weather, so we can prevent our children from getting wet. Children today take pain relievers when they have headaches. They have all kinds of toys (take a look at any toy store), and often take their toys for granted within a few hours or days.

Now let's compare my life with today's children. They don't get enjoyment from riding in a vehicle to get to their destination. They don't enjoy arriving home dry on a rainy day. Technology has made children's lives a lot easier by alleviating the suffering children faced in the past. A question always exists in the back of my mind, "Was I less happy than today's children?"

As I discussed in the chapter on happiness, quality does not matter. Instead, it is a question of how much we suffer—the more we suffer, the more we enjoy.

I used to suffer more from headaches when I was a child and enjoyed more when the pain was gone. Advances in science have given us pain relievers, so we don't suffer as much from headaches, obviously less enjoyment when the pain is gone. When you are a guest in my home and you have a headache, I don't know whether to offer you pain relief or not. If you take pain relief immediately when the headache starts, you'll suffer and enjoy it a little, but if you don't take the pain relief immediately, you'll suffer more and enjoy more when the pain is gone.

It's true that technology has made our lives easier, but is it true that it has brought us happiness? Technology has improved the quality of our lives but has it improved the level of our happiness? Technology has made it possible for us to suffer less, but at the same time it has also made us enjoy less.

Does technology bring us lifetime happiness?

The air conditioner in my home is broken, which means I suffer on hot days. Should I fix the air conditioner? When I suffer on hot days, then I really enjoy the cool evenings; but if I fix the air conditioner, I won't suffer from the heat as much nor enjoy the cool evenings as much.

We all take the comforts provided by technology and science for granted. Experiencing new technologies gives us temporary happiness, but sooner or later we get used to them. Then the only time we enjoy them is when we first suffer by missing them. Once again, the total enjoyment equals the total suffering experienced when we missed them. For instance, we all take electricity for granted. Power outages make us suffer; thus, we enjoy electricity when it is turned back on. As soon as our enjoyment equals our previous suffering, we take electricity for granted once again. The longer power outage causes more suffering, consequently more enjoyment.

Conclusion:

Technology may give us temporary joy, but believing that technology brings us lifetime happiness is a false belief. [F22]

Chapter 24

SUICIDE AND BALANCE

Why does a person kill himself? When a person faces significant problems, losing his balance and suffering severely to the extent that he cannot get his balance back, he attempts suicide.

When a factor affects us, we lose our balance, and we take action to regain our balance (fact two). When there is no motion to make in order to regain our balance, we suffer until we adapt to the new situation. For instance, when we lose our balance because of the death of a loved one, there is no motion that can help us regain our balance. We suffer until we adapt to the new situation. In some cases, when a person doesn't adapt to the new situation, he may attempt suicide.

Conclusion:
Suicide takes place when one cannot regain his or her balance.

Chapter 25

IS THE HUMAN A PERFECT CREATURE?

My jealousy bothers my friends and myself. People don't like me because I'm selfish. I'm not successful because I lack self-confidence. I don't have a good job because I'm not smart. I'm not proud of myself because I can't accomplish it. I can't have my own business because I'm not a risk-taker. I don't have long-term friends because I'm not flexible and people keep their distance from me because I'm a pessimist.

Nobody gets close to me because of my stinky breath and smelly armpits. I can't take off my shoes in others' presence because of my feet's stench. I suffer from hair loss and being hard of hearing. Bad teeth and blurred vision bother me…

Conclusion:

I have numerous abnormal personalities and physical characteristics. **Should I consider myself a perfect creature?**

Chapter 26

CAN WE SUCCESSFULLY CHANGE OTHERS?

Fact seven- the degrees of our personality, physicality, and mental and environmental conditions determine the levels of our wants and fears, and the sum of the outcomes of those wants and fears determine our motions.

Fact seven indicates that environment, personality, physicality, and mental condition determine our behavior. The behavior will change if we modify these factors.

ENVIRONMENT

Example 1- An undisciplined person behaves differently when he or she joins the Army.

Example 2- When a peasant moves to a big and compact city from a village, his or her behavior changes from calm to hectic.

Example 3- An honest person behaves dishonestly when associating with dishonest people.

PERSONALITY

Example 1- A trusting person becomes distrustful when he or she doesn't receive the lent money.
It should be noted that it takes a short time to change from trusting to distrustful, but it takes a longer time to change from distrustful to trusting.

Example 2- An insecure person behaves differently when her living condition changes from insecure to secure.

PHYSICALITY

Example 1- You move a coffee table differently after you start developing back pain.

Example 2- You behave nervously when you get hard of hearing.

Example 3- Your eating behavior changes when your blood cholesterol level rises.

MENTAL CONDITION

Example 1- You behave sad and hopeless when you become depressed.

Example 2- Lack of confidence in your appearance makes you behave differently.

Example 3- Confidence in your knowledge makes you speak with confidence.

When people's behavior is changed, are these changes permanent or temporary?

Some behavior changes are permanent, and some are temporary. There are two methods of behavior-changing:

Method I- Increasing or decreasing wants and fear.
Method 2- Modifying the character genes.

Method I- Increasing or decreasing wants and fear to change behavior.

Changing others' behavior by increasing or decreasing want and fear is a normal practice among people. Mothers promise to fulfill their children's wants in exchange for good behavior. For instance, mothers tell their children that if they behave good, they buy them the toy they wanted. An example of fear will be when the wives threaten their husbands with divorce if they don't quit coming home late.

Increasing and decreasing wants and fear usually lead to temporary behavior change, and people return to their previous behaviors when the fear or want is over. For instance, some people with low degree discipline characteristics return to their previous discipline when released from the Army.

The above review is a general analysis of behavior change. To change one's behavior, we need to study and examine factors related to each individual and find a specific training to modify the person's related behavior, which is not the aim of this book.

Method 2- Modifying the character genes to change behavior.

The individual's personality or character (chemical setting of the related gene) plays a major role in behavior. For example, a person with a personality that lacks discipline will behave undisciplined when he or she is released from the army. A person with a high degree of discipline characteristic will remain disciplined when released from the army.

Disciplinary training is more effective for children than adults. As such, the effect of army disciplinary training on a person with a low degree of discipline is temporary.

The question is, how can we change people's characters permanently? People's characters change slightly with long-term training. We must modify the related genes if we want a quick and permanent change.

Is it possible to modify the character's genes? The following video clips address this question.

Gene editing:
www.youtube.com/watch?v=NlCoDbmwFVs

CRISPR: Gene editing and beyond - YouTube

Method of the Year 2011: Gene-editing nucleases - by Nature Video - YouTube

Sixty years ago, around 1960, it was difficult to envision that we could change peoples' look and figures, but now, cosmetic surgeries are common, and we can easily change people's appearance.

Gene editing is a new science that is at an early age. I'm sure that we will soon develop technologies to modify personality genes and change people's behaviors.

Psychologists and psychotherapists barely succeed in changing the patients' behaviors, and their success depends on the patient's personality.

I believe that sooner or later, psychologists, psychotherapists, and psychiatrists' clinics will consider genome modification.

Conclusion

Behaviors can be changed by training, which takes a long time, and the change may not be definite.

Genome editing is a new technology that could definitely change human characteristics.

Chapter 27

ANALYSIS OF EMOTIONS

Emotions are feelings such as love, anger, hate, joy, envy, worry, sadness and so on. These feelings are the result of certain chemical processes in our brains. We think there is something beyond the body called emotion, but there is nothing beyond the body; there is only the reflection of chemical process that make us feel sad, happy, worried, or angry.

Consumption of alcoholic beverages makes us feel either happy or sad due to the chemical processes in our bodies. Using drugs and smoking marijuana changes our mood by creating certain chemical transactions in our brains. Pregnancy hormones totally change women's feelings and puberty hormones make us fall in love, or to feel jealousy towards our beloveds. Listening to poetry or music generates specific chemicals in our brains that result in specific emotional reactions. The psychotherapists' words generate the right chemicals in the brain of a patient towards patients' better feelings. The lack or excess of some hormones causes certain abnormal feelings. Anxiety can be cured with medication and there is medication for depression. Manic-depressives can become normal by taking simple medicines. So, abnormal feelings are the result of chemical formations that are other than normal. Abnormal feelings can go back to normal through the consumption of the right chemical formula.

✻✻✻✻✻✻✻✻✻✻✻✻✻✻✻✻✻✻✻✻

Every day, researchers are learning more about the chemicals that the neurons (NUR-ons) in the human brain use to communicate with each other. They now know that all the feelings and emotions that people experience are produced through chemical changes in the brain. The "rush" of happiness that a person feels at getting a good grade on a test, winning the lottery, or reuniting with a loved one occur through complex chemical processes. So are emotions, such as sadness, grief, and stress. When the brain tells the body to do something, such as to sit down or run, this also sets a chemical process in motion. These "chemical communicators," or neurotransmitters, are the "words" that make up the language of the brain and the entire nervous system.

****

"THERE ARE MEDICATIONS TODAY THAT MAY

HELP IN ALMOST ANY EMOTIONALLY CONNECTED ISSUE: OBSESSIVE WORRY, PANIC ANXIETY, GENERALIZED ANXIETY, RESTLESS LEGS, TOO MUCH WEIGHT, TOO LITTLE WEIGHT, INATTENTION, DECREASED MEMORY, DRUG ADDICTION, MANIC HIGHS, DEPRESSIVE LOWS, PREMATURE EJACULATION, IMPOTENCE, DECREASED SEX DRIVE, TOO MUCH SEX DRIVE, PMS, ANGER, INSOMNIA, PAIN, PSYCHOSIS, AND MANY MORE. THE TOOLS ARE POWERFUL. THEY MUST BE USED IN AN APPROPRIATE, SENSITIVE MANNER.

"THE MIND AND THE BODY ARE ONE AND SHOULD NOT BE TREATED SEPARATELY"

THE STUDY OF THE BRAIN MAY BE THE LAST AND MOST EXCITING BIOLOGICAL FRONTIER. RESEARCH IS INTENSE, AND THE FINDINGS ARE HIGHLY ENCOURAGING TO THE UNDERSTANDING OF DEPRESSION, ANXIETY, PANIC, OBSESSIVE WORRIES, WEIGHT ISSUES, INATTENTION PROBLEMS, PAIN, MEMORY, ANGER, ENERGY, SLEEPLESSNESS, LOGIC, MOTIVATION, ASSOCIATION, DRUG ABUSE, AND SEXUAL ISSUES, TO NAME JUST A FEW. FROM RESEARCH WE KNOW THAT CHEMICALS IN THE BRAIN (NEUROTRANSMITTERS) PLAY A LARGE PART IN THE SYMPTOMS LISTED ABOVE. PSYCHIATRIC MEDICATIONS TODAY CAN OFTEN HELP

SIGNIFICANTLY IN ABATING ISSUES SUCH AS THOSE LISTED ABOVE.

IF DOPAMINE IS ALTERED, THEN ONE MAY BE OUT OF TOUCH WITH REALITY; IF GABA IS ALTERED, ONE MAY BE ANXIOUS; IF ACETYLCHOLINE IS ALTERED, THEN ONE MAY HAVE MEMORY PROBLEMS; IF NOREPINEPHRINE IS ALTERED, ONE MAY BE MANIC HIGH; AND IF SEROTONIN IS ALTERED, ONE MAY BE DEPRESSED, WORRIED, MORE IRRITABLE, AND HAVE INSOMNIA. TODAY WE NOT ONLY KNOW WHICH NEUROTRANSMITTERS PRODUCE CERTAIN EMOTIONS, WE EVEN KNOW ABOUT NEUROTRANSMITTER SUBSYSTEMS. FOR EXAMPLE, THERE ARE AT LEAST EIGHTEEN SUBTYPES OF SEROTONIN RECEPTORS ALONE. WE EVEN KNOW WHAT THE SUBNEURO TRANSMITTERS DO, AND WE HAVE MEDICATIONS THAT WILL NOT ONLY AFFECT SPECIFIC NEUROTRANSMITTERS BUT ALSO SUBSYSTEMS. THE BIOCHEMISTRY OF EMOTIONS HAS BECOME A SCIENCE IN AND OF ITSELF."

THE MINIRTH CLINIC P.A.

****

When we hear something funny, a chemical transaction causes us to lose our balance. We perform a motion through laughter to regain our balance (chapter one).

When we get angry, the production of certain chemicals in our brains causes us to lose our balance and we want to make a motion to regain it.

When we hear an unhappy story, the chemical in our brains makes us feel sad. We lose our balance by the sadness feeling and we make a motion by crying to regain our balance (chapter one).

Brain chemicals and human actions
https://www.youtube.com/watch?v=09eVouoCLaw

There is nothing beyond the body such as emotion. Emotional feelings are the results of chemical processes in the body and brain.

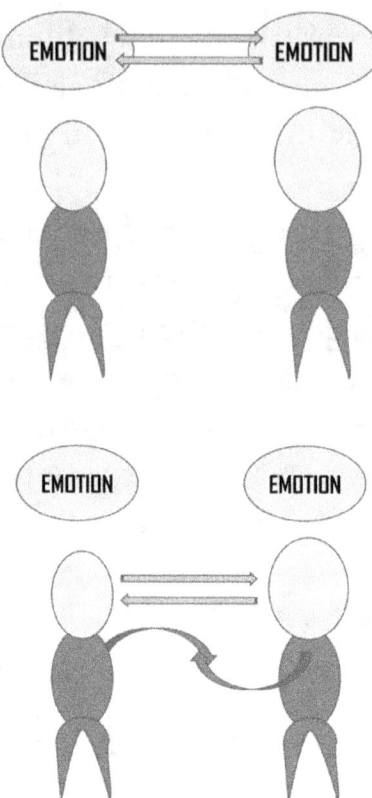

Emotional connections are the results of certain chemical
processes in our brain

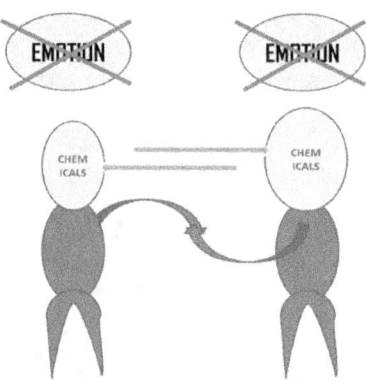

Watch these videos to learn about the functions of neurotransmitters, hormones, and chemicals:

Neurotransmitters
Stimulating serotonin release - Medication for Depression
http://www.youtube.com/watch?v=qMsWtP3VQ

Testosterone, Estrogen
https://www.youtube.com/watch?v=DGVPQF03tfl

Dopamine
https://www.youtube.com/watch?v=o2T-7_g6yUU

Emotional connections between humans are a series of chemical transactions in the brain. Take falling in love as an example; when we see a nice person of the opposite sex, we start loving her/him due to the chemical process in our brain.

We receive signals through the eyes and ears such as the beauty of a woman or the pleasant words of a man; these signals through our eyes and ears transfer to the neurotransmitters in our brain and the production of certain chemicals make us start to love the person or fall in love with her or him.

When somebody calls you a bad name, you get angry; this anger as an emotional feeling is the result of certain chemical processes in your brain produced by neurotransmitters via the signals received by your ears and eyes.

There is nothing beyond our bodies such as emotions. Emotional feelings are the results of chemical processes in our brains.

We receive signals through our eyes, ears, nose, skin, and tongue. The reflected feelings are the chemical process. These chemical transactions through neurotransmitters give us the feeling of sadness, cheer, hate, anger, revenge, contentment, love, and so on.

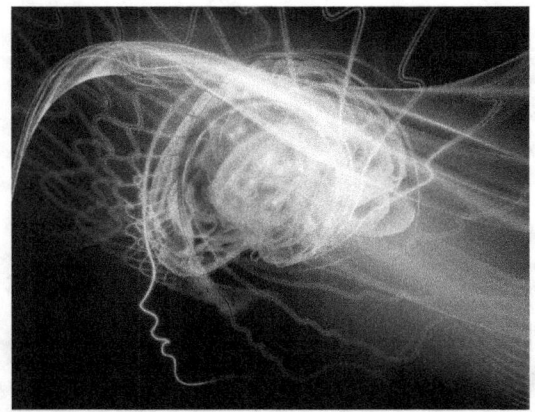

Examples: Signals from the eyes
    See a handsome man or a beautiful woman
    See an injured person covered in blood
    See the ocean
    See your car's flat tire

Examples: Signals from the ears
    I love you
    Gossip
    You are called a bad name
    Compliment

Examples: Signals from the nose
    Smell a flower
    Smell your baby
    Smell someone's bad breath

Examples: Signals from touch(skin)
    Someone touches your back in sympathy
    You touch the opposite sex
    Someone slaps you

Examples: Signals from the tongue
 Tasting sugar
 Tasting vinegar

Conclusion:

Emotional feelings are the results of the certain chemical processes in the brain. There is nothing beyond the brain.

Chapter 28

WHAT IS THE SOUL?

A- What Is Life?
B- What Is the Soul?
C- What Is Psyche?
D- What Is Spirit?
E- What Is Emotion?
F- What Is Mentality?
G- What Is Mental Disorder?
H- What is I? What am I? What is the definition of I?

A- What is life? B - What is the soul?

When we say I, to what do we refer?

Examples:
I am the only one who can stand you.
Nobody is as patient as I am.
I do what I want.
I forgive you.
I never accept unreasonable orders.
I don't take no for an answer.
Am I just this body and brain or is there also something beyond this body called soul?

What is the soul?

What is the definition of the soul? I've asked this question of many people and nobody has yet been able to describe the soul. Some say the soul is something, but we don't know what. How can we believe in things if we don't know what it is? Others believe that soul is energy. Here are the energies in nature:

- Heat/thermal
- Light
- Chemical
- Electrical
- Mechanical/potential

These energies have a material base.
E=MC2, the most famous Einstein formula, tells us that matter can convert to energy. Thus, the soul is material if we believe that soul is energy. We can measure the five types of energies. Can we measure the soul? If we can't measure things, it means they don't exist.

Recent discoveries about the soul:

http://www.youtube.com/watch?v=aRzrYNVXF28

http://www.youtube.com/watch?v=6qsIK4zEbT4

Craig Venter's creation of synthetic life engaged our minds with a major question: What is a soul when we can produce life in a laboratory?

The following is the report by The Guardian's

Science dated May 20, 2010:

****

Craig Venter and his team have built the genome of a bacterium from scratch and incorporated it into a cell to make what they call the world's first synthetic life form.
Scientists have created the world's first synthetic life form in a landmark experiment that paves the way for designer organisms that are built rather than evolved.

The controversial feat, which has occupied 20 scientists for more than 10 years at an estimated cost of $40m, was described by one researcher as "a defining moment in biology".

Craig Venter, the pioneering US geneticist behind the experiment, said the achievement heralds the dawn of a new era in which new life is made to benefit humanity, starting with bacteria that churn out biofuels, soak up carbon dioxide from the atmosphere and even manufacture vaccines.

However critics, including some religious groups, condemned the work, with one organization warning that artificial organisms could escape into the wild and cause environmental havoc or be turned into biological weapons. Others said Venter was playing God.

The new organism is based on an existing bacterium that causes mastitis in goats, but at its core is an entirely synthetic genome that was constructed from chemicals in the laboratory.

The single-celled organism has four "watermarks" written into its DNA to identify it as synthetic and help trace its descendants back to their creator, should they go astray.

"We were ecstatic when the cells booted up with all the watermarks in place," Dr Venter told the Guardian. "It's a living species now, part of our planet's inventory of life."

Dr Venter's team developed a new code based on the four letters of the genetic code, G, T, C and A, that allowed them to draw on the whole alphabet, numbers and punctuation marks to write the watermarks. Anyone who cracks the code is invited to email an address written into the DNA.

The research is reported online today in the journal Science.
"This is an important step both scientifically and philosophically," Dr Venter told the journal. "It has

certainly changed my views of definitions of life and how life works."

The team now plans to use the synthetic organism to work out the minimum number of genes needed for life to exist. From this, new microorganisms could be made by bolting on additional genes to produce useful chemicals, break down pollutants, or produce proteins for use in vaccines.

Julian Savulescu, professor of practical ethics at Oxford University, said: "Venter is creaking open the most profound door in humanity's history, potentially peeking into its destiny. He is not merely copying life artificially ... or modifying it radically by genetic engineering. He is going towards the role of a god: creating artificial life that could never have existed naturally."
This is "a defining moment in the history of biology and biotechnology", Mark Bedau, a philosopher at Reed College in Portland, Oregon, told Science.
Dr Venter became a controversial figure in the 1990s when he pitted his former company, Celera Genomics, against the publicly funded effort to sequence the human genome, the Human Genome Project. Venter had already applied for patents on more than 300 genes, raising concerns that the company might claim intellectual rights to the building blocks of life.

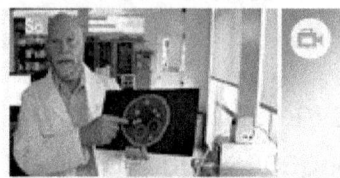

Craig Venter explains synthetic genomics
Creator of first synthetic genome describes how
replacing the chromosomal software of a cell
transforms its protein hardware

Genome editing

Error! Hyperlink reference not valid.www.youtube.com/watch?v=NICoDbmwFVs

CRISPR: Gene editing and beyond - YouTube

Method of the Year 2011: Gene-editing nucleases - by
Nature Video - YouTube

Do we have a soul
Synthetic Biology – Inventing the Future
http://www.youtube.com/watch?v=iRO0-fMIW9I

C- What is the psyche?
    I have no idea, what is the psyche, do you? Can

you describe the psyche?

D- What is spirit?

Can you describe spirit? Neither can I.

Let's dig further into the soul to find out more about it.

What are the duties and tasks of the soul? Can we say any of the following is the soul's task?

1. **Do our lives depend on the soul?**
2. **To love, to feel, and to express feelings?**
3. **To learn and memorize?**
4. **To think?**
5. **Does conscience depend on the soul?**
6. **Does measuring, evaluating, and recognition depend on the soul?**

1- Does our life depend on the soul?

The above studies and the creation of synthetic life prove that life involves only a certain chemical setting, and we can produce it in the laboratory.

2- To love, to feel, and express feelings?

The following video tells us that emotions are chemical transactions in the brain.

http://www.youtube.com/watch?v=RwKMOpjsGCl

testosterone
https://www.youtube.com/watch?v=DGVPQF03tfl

How are we connected by love?

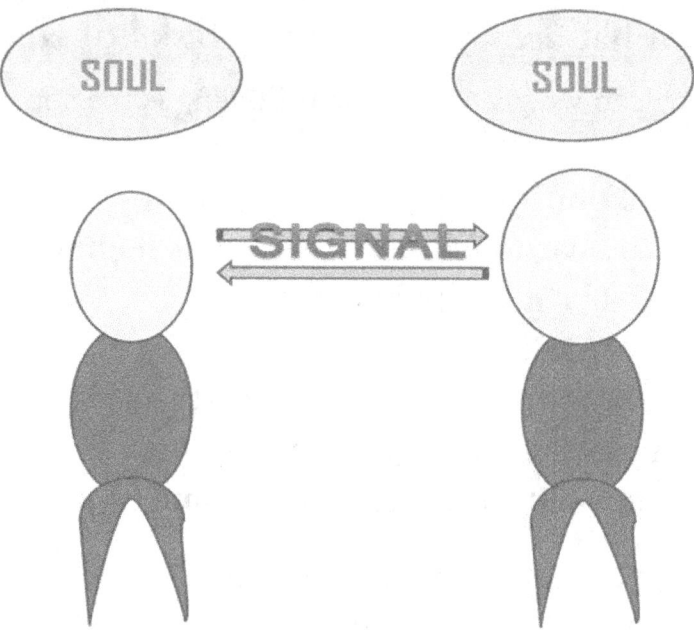

Love as something beyond our bodies is an untrue belief. Love is a feeling, and feeling is a chemical process in the brain. There is nothing beyond our bodies, such as love or feelings. When we meet a person of the opposite sex, the image and words of the person received through our eyes and ears send signals to the neurotransmitters in our brains, and the production of

certain chemicals causes us to like the person, and that's
how we fall in love.

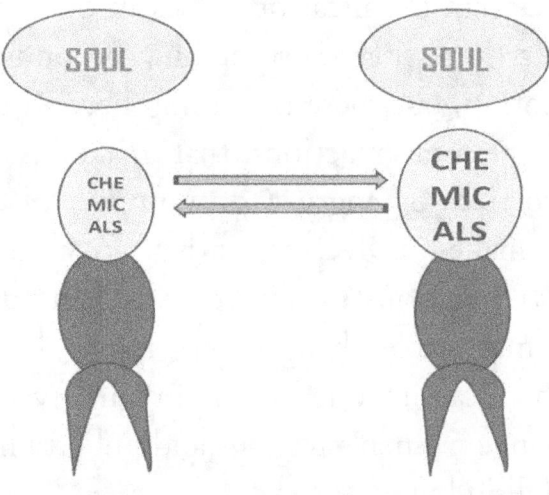

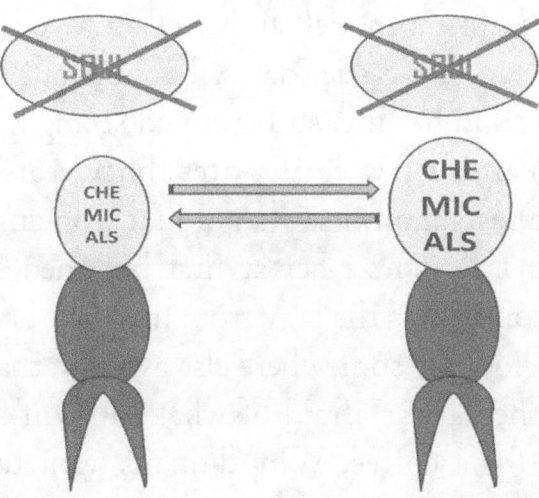

Emotions are feelings such as love, anger, hate, joy, envy, worry, sadness and so on. These feelings are the result of certain chemical processes in our bodies and brains. We think there is something beyond the body called emotion, but there is nothing beyond the body; it's only chemical reactions that make us feel sad, happy, worried, or angry. Consumption of alcoholic beverages makes us feel either happy or sad due to chemical transactions in our bodies. Using drugs and smoking marijuana changes our mood by creating certain chemicals in our brains. Pregnancy hormones totally change women's feelings and puberty hormones make us fall in love.

MY FAVORITE WOMAN

I met a lady about 38 years old, who was beautiful, with a body like a model, and I fell in love at the first sight. After a few casual dates, I found out there was another man in her life, so I stopped seeing her and we lost touch after a few years. After thirty years, as I was eating in a restaurant, I saw an old fat woman watching me. I didn't recognize her at first but then I realized that the woman was the lady from my past. I turned my eyes away, looking somewhere else as a sign that I didn't recognize her. I was afraid of what to say if she asked me to go for a coffee. Why didn't I want to see this lady who had once been my dream girl?

Obviously, the fat face with the wrinkles didn't send the same signal as 30 years ago to my brain and the related chemicals were not produced. If we believe that we are connected by our souls when we fall in love, why didn't this connection happen this time? My soul and her soul are still the same. Why did these two souls connect 30 years ago but not after 30 years had passed? I have no doubt that if she had looked the same as before, my feelings would be returned after 30 years. My point is that there is nothing beyond our bodies such as soul and love; it's the chemical transaction in our brains that gives us feelings.

3- Learning and memorizing

We memorize information through the chemical transactions in the brain. The following video shows how our memories work and no scientific evidence tells us that the soul exists or that our memories depend on the soul.

http://www.youtube.com/watch?v=VYNh3-21HmA

I have had five surgeries and each time the doctor administered drugs that rendered me unconscious for the operation. My body did not feel anything, but my soul was conscious so I must remember all things during the operations, but I remember nothing. Do we still believe in the soul?

Watch this video about learning, memorizing and intuition

https://www.youtube.com/watch?v=1cdjUT4lg14

4- Thinking

Does our soul do the thinking or does our brain do the thinking? Read about this in detail in the chapter, What Is Thinking?

5- Does conscience depend on the soul?

Conscience is nothing but fear, and fear is a chemical transaction in the brain.

Example 1: You are the guardian of an orphan. Will you feel bad if you spend his or her money on yourself? Is the feeling moral or is it the fear of suffering?

Example 2: You are honest with your wife and now have to lie to her because you've cheated on her. You have a bad feeling if you do so. Is this feeling moral or is it a fear of suffering? Read more about this in the chapter, Is Morality a Human Value?

6- Measuring, evaluation, and recognition.

We can measure things based on past experiences that we've learned and memorized in our brain and the same can be said for our evaluation and recognition.

https://www.youtube.com/watch?v=dKaqFz_WoIw

Our brain learns how to measure, evaluate, and recognize bad and good through experimentation.

https://www.youtube.com/watch?v=iMYJn2-1u2Q

Conclusion:
How do we believe in the soul when we can't describe it and it has no duties or tasks? Life, to feel, respond to feeling, learning, memorizing, conscience, measuring, evaluating, and recognition are all brain activities and tasks; there is nothing external. [F18]

E- What is Emotion

Emotions are nothing but chemical transactions in the brain. Read more about emotion in the chapter on Emotion.

F- What is Mentality?

There is nothing beyond our bodies such as mentality. The mentality is the processed series of chemicals in our brain. Read more about mentality in the chapters on Is Morality a Human Value? and, Is It Mental Disorder or Chemical Disorder?

G- What is Mental Disorder?

There is nothing beyond our bodies such as the

mental or a mental disorder. A mental disorder is just a certain chemical formation in the brain which differs from the normal chemical formation. Read more in the chapter on, Is It Mental Disorder or Chemical Disorder?

H- What is I? What am I? What is the definition of I?

Am I the body and the brain plus the soul or am I just the body and the brain? Read more in the chapter on, What Are We?

Conclusion:

The scientific discoveries in the above videos and the results of the previous logical discussions prove that our actions, reactions, feelings, and responses to feelings are all chemical processes in our brains, and there is nothing external.

Chapter 29

WHAT IS THINKING?

How Do We Think?

Thinking is the mathematical calculation and evaluation of the brain to find answers to questions, based on the information stored in the memory. In some cases, thinking is the brain's evaluation of joy or suffering.

Calculation and Evaluation

A brain's functionality is similar to that of a computer. Both have memory storage and a processor to evaluate the new information. If you ask a very young boy how much two plus five is, he can't answer, but when he learns numbers, quantities, and the basics of calculation, he can calculate based on the information stored in his brain. The basic processor already exists in the child's brain, but he can't answer the question because of the lack of information. That's the same with computers; if you enter 2+5 in a basic computer, it will not answer but when you download calculator software, you'll get the answer based on the stored information.

This is the same scenario for adults. For instance, if you are new in the country and want to buy insurance for your car, you will ask yourself which insurance company and at what rate. Obviously, your brain has no answer, so you call a friend and search online to collect information, then your brain has an answer based on the recorded information. Your brain might not give you the best answer if you didn't give it complete information, for example, if you didn't search enough to find the promotion from company x, it's obvious that your brain can't give the best answer because of lack of information.

Our brain always gives the wrong answers because of a lack of information. A normal and healthy brain never makes mistakes in calculation or evaluation; it's the lack of information or incorrect inputs that result in wrong answers.

Every brain has a specific processor with a certain speed. Smart people have a high-speed processor and can retrieve all the information related to a question that is stored in memory, evaluate the information, calculate instantly and have the answer quickly. Regular people's processor operates at normal speed and may not be able to retrieve all the information related to a question, so the answer may take a little longer and may not be as satisfactory as that of a smart person because not all the information is retrieved. Dumb people have a slow processor and miss retrieving most of the information or there is a problem with the memory functionality or capacity. Mentally retarded people have a damaged processor or memory dysfunction.

It's not proper to classify people into four groups smart, regular, dumb, and mentally retarded. Every person has a processor with a certain working speed and a memory with specific functionality. We may evaluate smartness from 0 to 100. The most mentally retarded will have 0 smartness and the very smartest will have 100 degrees of smartness, so every person has a certain degree of smartness between 0 and 100.

B- Evaluation of Joy and Suffering

There is a constant evaluation of joy and suffering in our brain, and every single action is the outcome of evaluated joy and suffering sums. We perform motions where there is more enjoyment or less suffering, depending on the situation.

****

HOW DOES THE BRAIN THINK?

By tadakaluri | Posted April 13, 2014

The human brain is the most mysterious and complex entity in the known universe.

It's a computer, a thinking machine, a fatty pink organ, and a vast collection of neurons - but how does it actually work? The human brain is amazingly complex - in fact, more complex than anything in the known universe. The human brain effortlessly consumes power, stores memories, processes thoughts, and reacts to danger.

In the most basic sense, our brain is the center of all input and outputs in the human body. Dr Paula Tallal, a co-director of neuroscience at Rutgers University, says the brain is constantly processing sensory information - even from infancy. "It's easiest to think of the brain in terms of inputs and outputs", says Tallal. "Inputs are sensory information, outputs are how our brain organizes that information and controls our motor systems'.

According to Dr Robert Melillo, a neurologist and the founder of the Brain Balance Centers, the brain actually predetermines actions and calculates the results about a half-second before performing them (or even faster in some cases). This means, when you reach out to open a door, your brain has already predetermined how to move your elbow and clasp your hand - maybe even simulated this movement more than once, before you even perform the action.

CNN Mobile apps

http://ireport.cnn.com/docs/DOC-1119917

Conclusion:

Thinking is the brain activity involved in evaluating information. Thinking does not depend on anything external. [F19]

Chapter 30

WHAT IS THE MIND?

Is the mind the same as the brain or is it something beyond the body? Can we touch or weigh the mind like the brain or is it not touchable?

How do you define the mind if you are asked to describe the mind?

When you ask people what they have on their mind, do you mean what they have in the brain or what they have beyond the body that is more powerful than the brain and supersedes the brain?

Researches on these videos tell us that mind and the brain are the same.

https://www.youtube.com/watch?v=esPRsT-lmw8

https://www.youtube.com/watch?v=Rl2LwnaUA-k&nohtml5=False

Conclusion:
Mind and brain are the same and there is nothing beyond the body.

Chapter 31

Every single human motion is based on the following facts

RAFII'S BASIC SEVEN FACTS OF MOTIONS

Fact 1- We always want to be in balance.

Fact 2- We make motions only when we lose our balance.

Fact 3- Loss of balance always generates want or fear in us.

Fact 4- Suffering and joy always accompany every motion.

Fact 5- More suffering leads to more enjoyment.

Fact 6- The sum of the outcome of our want and fear determines our motions.

Fact 7- The degrees of our personality and physical characteristics, and mental and environmental conditions, determine the levels of our wants and fears on every motion, and the sum of the outcomes of those wants and fears determine our motions.

Chapter 32

WHAT ARE WE?

Let's remember the things that you were doing a few hours ago: anything, such as taking a shower, putting on makeup, shaving, going to the bank, visiting a friend, arguing with your employer or employee, laughing, crying, or locking your car. Place yourself back in that time and imagine that all the conditions for any of these actions remained the same. Conditions are including surrounding condition; environmental condition, your mental and physical condition, and your personality condition. You would repeat each of the motions exactly as they occurred before and return to this precise moment with all the same exact conditions surrounding you as they do now. It would be the same for your motions yesterday, last month, last year, and all the way back to when you entered this world. So, if you were to go back to the time that you were born and if all conditions remained the same, you would perform the same motions one after another up to this moment. As a matter fact the properties of atoms and their tend to transact with each other, create the nature regulatory.

Now, what would happen in the future? Can we

predict the future? Let's start with a few simple examples.

Example one:

You buy a diamond ring for your wife or girlfriend, and you know that she will be happy and she will kiss you. This is your prediction of the future. You give her the ring, and she becomes happy and kisses you as predicted.

Example two:

You buy a remote-controlled car for your five-year-old son. You know that he will be very happy and will enjoy playing with it for hours. You give the toy to your son, and he is very happy and plays with it for hours. It happens as you predicted. Your prediction is based on your knowledge of your son's personality and desire, as well as on the experiment of seeing your son like such a toy, lose his balance, and want to play with it.

Example three:

You buy a remote-controlled car and give it to your five-year-old nephew, expecting he will be very happy; however, this does not happen. His mom tells you that he has a car exactly like it. Your prediction was wrong because you didn't have the knowledge of the condition. If you knew that he already had a car like it, you would not expect him to be happy, and you would have predicted the outcome correctly.

Example Four:

Your daughter marries a man that you know, and you predict that she will be happy based on your knowledge of their personalities. Most of the time the outcomes of our predictions are wrong because we don't have the knowledge of other conditions such as, bad friends, the personality of the mother-in-law, future economy, and hundreds of other factors. Before we dig deeper into future human actions, let's think about other things in nature.

You are looking at an ant in your backyard, following it to see what it's doing. It picks up a seed and goes to its nest. Let's imagine that you have the tools to mark one of the atoms in the seed to follow where it goes in order to see what happens. Let's say you picked a carbon atom. The seed will be stored under the ground in the ant's nest. Later, another ant eats part of the seed and the atom enters the body of the ant. The atom then enters the soil from the discharge of the ant and stays there for years. One day the atom enters the root of a plant and moves to the surface of a leaf and into the atmosphere through evaporation. It travels in the air to many places and later, in a drop of rain, lands on a mountain and goes underground, traveling back to the surface of a stream and continues moving.

Is the movement of the atom an accident or it is based on the regularities of nature? To analyze these complicated movements, let's look at a few simple examples.

Example A:

When you put oxygen and hydrogen together under certain conditions, they combine and become water. If you repeat this experiment hundreds of times under the same conditions, you will, without exception, get water. Each atom in this universe has specific properties that determine its tendency to combine with other atoms.

Example B:

You are in front of a pool table and only the cue ball remains on the table. You hit the ball toward the side of the table with a specific angle. The ball hits one side of the table, then rolls and hits the other side and stops in a certain location. If you place back the ball in the exact same position as before and hit the ball again with the exact same force and angle, the ball would make the same exact movements and stop in the same exact location. In reality, it is almost impossible to hit the ball with the exact same angle and force; even if we were able to do so, the result may vary slightly because of other factors, such as temperature, air circulation in the room, or variances in the texture of the surface of the ball.

Example C:

You are in the middle of a pool game, and it's your turn to hit the ball. There are numerous balls on the table. You hit the cue ball, and it hits a second ball; the second ball hits a third ball, and then the cue ball hits the side of the table, hitting the ninth ball. In this scenario, four balls moved and stopped in certain locations. If we put the balls back in the same location and hit the cue ball again with the exact same force and angle, the same four balls would move and stop in the same exact locations (physic laws).

Watch these video clips:

Pool table-amazing

http://www.youtube.com/watch?v=olQO5IHi8KQ

http://www.youtube.com/watch?v=7GA3ySz4el4

Example D:

You are playing with a deck of fifty-two cards in your hand. You shuffle the cards. We don't know what the first card is, but that particular card is not there by accident. If you were to place every single card back in the exact positions it had before and shuffle exactly as you did earlier, not only the first card, but every single other card would be in the exact same location as the first time.

Example E: You are in front of a lottery cage that contains ten balls marked from one to ten. You memorize the locations of each ball and then give the cage one spin. Each ball makes several movements and ends up in a new location. These new locations do not occur by accident. The resultant of all forces on each ball from the inner surface of the cage and the other balls creates a chain of movements that caused the ball to move to a precise location and position. If you could place each ball in its exact original location and position, turn the cage at the same speed and under the same conditions, the new position of each ball would be the same. It would be virtually impossible to place each ball back in the exact same location and turn the cage with the same conditions; moreover, even if we could repeat all the same conditions, we would also have to maintain additional factors, such as air circulation, temperature, air density, and any other factor that may have affected the original movement of the balls.

It's possible that one day we will create a computer program or other device to predict the winning numbers if we could supply it with all information and conditions.

Indeed, the whole universe is like a giant lottery cage, and each atom is like a ball in the universe. The properties of each atom and the existence of the laws of nature determine every single movement of each atom. Focusing on the chart of elements leads us to the basis of nature regulations.

Periodic Table of Elements

1 H 1.008	2 IIA 2A											13 III A 3A	14 IV A 4A	15 VA 5A	16 VI A 6A	17 VII A 7A	2 4.003
3 Li 6.968	4 Be 9.012											5 B 10.81	6 C 12.01	7 N 14.01	8 O 16.00	9 F 19.00	10 20.18
11 Na 22.99	12 Mg 24.31	3 IIIB 3B	4 IV B 4B	5 VB 5B	6 VI B 6B	7 VII B 7B	8 ← ← 8	9 VII I	10 → →	11 IB 1B	12 IIB 2B	13 Al 26.98	14 Si 28.09	15 P 30.97	16 S 32.07	17 Cl 35.45	18 39.95
19 K 39.10	20 Ca 40.08	21 Sc 44.96	22 Ti 47.87	23 V 50.94	24 Cr 52.00	25 Mn 54.94	26 Fe 55.85	27 Co 58.93	28 Ni 58.69	29 Cu 63.55	30 Zn 65.38	31 Ga 69.72	32 Ge 72.63	33 As 74.92	34 Se 78.96	35 Br 79.90	36 83.80
37 Rb 85.47	38 Sr 87.62	39 Y 88.91	40 Zr 91.22	41 Nb 92.91	42 Mo 95.96	43 Tc (98)	44 Ru 101.1	45 Rh 102.9	46 Pd 106.4	47 Ag 107.9	48 Cd 112.4	49 In 114.8	50 Sn 118.7	51 Sb 121.8	52 Te 127.6	53 I 126.9	54 131.3
55 Cs 132.9	56 Ba 137.3	* 57-71	72 Hf 178.5	73 Ta 180.9	74 W 183.8	75 Re 186.2	76 Os 190.2	77 Ir 192.2	78 Pt 195.1	79 Au 197.0	80 Hg 200.6	81 Tl 204.4	82 Pb 207.2	83 Bi 209.0	84 Po (210)	85 At (210)	86 (222)
87 Fr (223)	88 Ra (226)	** 89-103	104 Rf (257)	105 Db (260)	106 Sg (263)	107 Bh (265)	108 Hs (265)	109 Mt (266)	110 Ds (271)	111 Rg (272)	112 Cn (277)	113 Uut --	114 Uuq (296)	115 Uup --	116 Uuh (298)	117 Uus --	118 --

242

	57 La 138.9	58 Ce 140.1	59 Pr 140.9	60 Nd 144.2	61 Pm (147)	62 Sm 150.4	63 Eu 152.0	64 Gd 157.3	65 Tb 158.9	66 Dy 162.5	67 Ho 164.9	68 Er 167.3	69 Tm 168.9	70 Yb 173.1	71 Lu 175.0
* Lanthanides															
** Actinides	89 Ac (227)	90 Th 232.0	91 Pa 231.0	92 U 238.0	93 Np (237)	94 Pu (242)	95 Am (243)	96 Cm (247)	97 Bk (247)	98 Cf (249)	99 Es (254)	100 Fm (253)	101 Md (256)	102 No (254)	103 Lr (25

Reference: IUPAC 2011 standards truncated to four significant figures. Sponsored Links

OptiView DAB detection Vivid signal intensity for IHC Customizable for each antibody www.VentanaOptiView.com

Bullet-Proof T-Shirts Boron-treated cotton T-Shirt can stop a bullet say scientists www.micmet.com

Chemistry help Chemistry 9th Edition Solutions. View Free! Cramster.com/Chang

The location and position of each atom in the universe at this very moment are not accidental. Instead, it is related to the previous locations of that particular atom and its connections with other atoms. As I'm writing this book, the temperature in my backyard is sixty-one degrees Fahrenheit. This temperature at this moment and location is not accidental but the result of the atmospheric conditions and other related factors from a second ago, an hour ago, yesterday, a month ago, last year, and million years ago. The temperature at this time tomorrow at this location will not be an accident, and it will be determined by today's factors. The same applies to the next day, week, year, and another million years into the future. It is not irrational to believe that humans will someday be able to forecast the weather weeks or even months in advance. Storms, floods, tornados, hail, and earthquakes will be predicted more accurately compared to today.

It seems that humans break these regulations through their actions, and things like global heat and atomic bombs change the atmospheric conditions. The changed direction of these conditions and human actions, result from nature's regularity mechanisms caused by the property of the atoms. Consider the examples below.

Example 1:

Nature knows what would be the temperature in my city tomorrow. If the broadcasting computers predict different temperatures, the computer received wrong information from the atmospheric condition.

Example 2:

You are watching a swimming competition, and you don't know who the winner is, but Nature knows.

Nature knows who is the strongest because of each participant's characteristics. If we created a computer with the ability to evaluate all factors involved and entered the detailed information of each participant, including physical characteristics, personality characteristics, and mental condition (DNA), along with the involved environmental factors, the computer would predict the winner.

Example 3:

You are taking care of an older man who will die in several days. You don't know when he will die, but Nature knows. If we entered all related information in the computer, such as physicality, personality, mental condition, and environmental conditions, the computer would tell us the exact time of his death. I have to add that the treatments he gets from you or his doctor and other medical services are all environmental conditions.

People argue in my seminars and personal discussions that we could change the future if we did certain things. I always explain that "IF" didn't exist at the time of the incident. For instance, in the above example, my father would still be alive if the nurse didn't fall asleep but gave my father medication on time. This "IF" couldn't exist. The surrounding environment and conditions made the nurse fall asleep. Alternatively, my father wouldn't die if the doctor didn't give him the wrong medication by lack of experience. "IF" is irrelevant, as the doctor would give the same medication if we could go back in time before the examination. The doctor would have no reason to prescribe different medications, as all factors involved in this scenario would be the same, including his knowledge and experience, the medicine producer advertising, his mental condition at the moment of examination, his personality (degrees of caring and responsibility), the surrounding environment, and other related conditions.

Reflecting on Hitler, for example, whom almost every reader of this book knows. "IF" Hitler didn't have that specific DNA, WWII wouldn't happen, but it was impossible to avoid Hitler's DNA, so IF didn't matter; thus, WWII would happen anyway.

WWII wouldn't happen "IF" Hitler grew up in a different environment, but that "IF" does not exist, and if we went back in time, Hitler would grow up again in the very same environment, and WWII would happen again.

Could we change the decisions of the US authorities who decided to drop the atomic bomb on Japan, if we went back in time and all the conditions remained the same? Obviously, no. Therefore, the explosion of the atomic bomb and change in the atmospheric condition was unavoidable. The decision of each person in the government committee was based on the regularity of nature and would be the same in the same condition, as it would be based on the regularity of the properties of elements. Therefore, if we could create an advanced computer with the ability to evaluate and analyze the movement of elements forming Hitler's parents and grandparents' DNA, including the surrounding environment, the computer would tell us that a child named Hitler with the specific DNA will be born and WWII will happen.

Moreover, the computer would predict the US decision, the explosion of the atomic bomb, and atmospheric change in the area. So "if" we could change Hitler's parents' or grandparents' DNA, we could prevent atmospheric change, but we could not avoid the change because the "if" couldn't exist. If we believe that the movement of elements in Hitler's parents' and grand parents' DNA was natural, then the US authorities' decisions were natural too based on their DNA and surrounding environment, and the movement of elements of the exploded atomic bomb that changed the atmospheric conditions were natural and unavoidable as well. In summary, if we went back to the time before Hitler was born and all the surrounding environment in the world would remain the same, Hitler would be born with the same DNA,

WWII would happen exactly as it happened, US authorities would make the same decisions, and the atomic bomb would explode and change the atmospheric condition.

Science is at an early stage in predicting the future. What I have in mind is like a navigation system. When you enter your destination address in your navigator, it tells you the exact time you'll arrive at your destination. This precise prediction is based on many related factors inputted into the navigator's computer, including the distance, number of traffic lights, number of stop signs, speed limits in residential streets, speed limits in the highways, and slow traffic or a traffic jam, and the computer output is based on the evaluated and calculated factors.

The location and position of each atom in
the Universe is not an accident

We can predict things in nature because every movement is based on the regularities and properties of the elements. Human's motions are the same. Every single motion is the result of the *wants* and *fears,* and the wants and fears are the results of a certain precise chemical process.

It's possible that someday, man will create an advanced computer or device to predict our motions. To have precise predictions, we have to enter much information into computers, including:

A- Personality characteristics: Our personalities have hundreds of characteristics, and we each have a certain degree of each characteristic, such as tolerance, envy, confidence, flexibility, generosity, responsibility, selfishness and so on. You may review the shortlist of personality characteristics in the chapter on Personality Characteristics of this book.

B- Physical characteristics: Age, height, weight, eye color, blood pressure, blood sugar level, cholesterol level, back pain, and hundreds of other factors.

C- Mental conditions: Depression, anxiety, inattention, insomnia, and other mental conditions.

As we discussed before, personality characteristics, physical characteristics, and mental conditions are all physical, and DNA includes the properties of the three characteristics. Now to shorten the sentences we use DNA instead of naming the three characteristics.

D- Environmental conditions: Weather, food, friends, society, culture, religion, job setting, education, economy, government, politics, and hundreds of other factors.

Our DNA and environmental conditions, determine the levels of our wants and fears. We act and react according to the sum of the outcome of our wants and fears.

Our DNA and environmental conditions determine the levels of our wants and fears, and the sum of the outcomes of those wants and fears determine our motions.

We could predict some human motions because they are based on regularities. If we are wrong in our prediction, it is because we failed to recognize all the related conditions and calculate the effects on that particular motion.

If we look deeper into nature and human actions, the following don't exist:

An accident doesn't exist:

Remember an injury you have experienced. If you

place yourself back to when the accident happened and if all the conditions remained the same, including your personality and physical characteristics and mental conditions and environmental conditions, the accident, and the injury would happen again. **[F26]**

Guilt and fault don't exist:

If we take the guilty person back to the moment of his or her crime and if all the conditions remain the same, including the personality and physical characteristics and mental and environmental conditions, the guilty person will act in the same way. Question: If we've swapped your DNA with that of Richard Ramirez, the serial killer, and you were to grow up in the exact same environment and conditions as he did, would you kill the same people or not?

You may think of a killer you know if you don't know Richard Ramirez. Imagine that we have a technology to replace your DNA with the DNA of a killer you know, at the time you were born, and you grew up in the same exact environment that he did, would you kill the same person or not? **[F27]**

A mistake doesn't exist:

Remember any mistake you have made. If you were to go back to the time of action and if all the conditions remained the same, including your personality characteristics, physical characteristics, mental conditions, and environmental conditions, you would perform the same action. [F28].

Conclusion:

The degrees of our personality and physical characteristics, and mental and environmental conditions, determine the levels of our sufferings and enjoyments on every motion, and the sum of the outcomes of those enjoyments and sufferings determine our motions. *What does that say about human nature?*

INDEX

REFERENCES

Brain chemicals and human actions   115
https://www.youtube.com/watch?v=09eVouoCLaw

Brain chemistry   117
Human illnesses,
http://www.humanillnesses.com/Behavioral-Health-A-Br/Brain-Chemistry-Neurochemistry.html

Brain chemistry   119, 127
The Minirth Clinic P.A.
http://www.minirthclinic.com/digest-downloadpg1.html

Brain – Can human build a brain?   142
https://www.youtube.com/watch?v=iMYJn2-1u2Q

Craig Venter and The Guardian's Science   137
http://www.guardian.co.uk/science/2010/may/20/craig-venter-synthetic-life-form

Dopamine   117, 121
https://www.youtube.com/watch?v=o2T-7_g6yUU

Estrogen   117
https://www.youtube.com/watch?v=DGVPQF03tfI

CONTACT INFORMATION

For further information or to share your experiences, please send e-mails to:

edrafii@gmail.com

www.eddierafii.com

.